Handbook of
ELECTRICAL HAZARDS *and*
ACCIDENTS

Edited by
Leslie A. Geddes

CRC Press
Boca Raton   New York   London   Tokyo

Library of Congress Cataloging-in-Publication Data

Handbook of electrical hazards and accidents / edited by Leslie A. Geddes.
      p.   cm.
  Includes bibliographical references and index.
  ISBN 0-8493-9431-7 (alk. paper)
  1. Electrical injuries. I. Geddes, L. A. (Leslie Alexander), 1921–
RA1091.H36   1995
617.1'22—dc20   95-17483
                                                                       CIP

No claim to original U.S. Government works
International Standard Book Number 0-8493-9431-7
Library of Congress Card Number 95-17483
Printed in the United States of America 1 2 3 4 5 6 7 8 9 0
Printed on acid-free paper

PREFACE

This book is written mainly for attorneys, physicians and investigators who are concerned with accidents associated with electric current. The most identifiable audiences consist of emergency medical personnel, biomedical engineers, manufacturers of medical devices, electric power companies, expert witnesses and accident investigators. The material is presented in two ways: non-technical and technical; the former is for attorneys and physicians, the latter is for their expert witnesses and engineers. The contents of this book are based on about fifty years of personal research and teaching medical and biomedical engineering students. Material has been included from experience as an expert witness for plaintiffs and defendants in many litigations involving electrical injury. There is no typical electrical accident, and it is hoped that the selected accidents described here and the material in the chapters will allow the reader to explain the cause of any particular electrical accident.

The first chapter presents a short history of electrical accidents, along with selected accident descriptions in the home, work place and hospital. Chapter 2 deals with the fundamental processes whereby electric current stimulates excitable tissue, such as sensory receptors, nerve and muscle. The response to the passage of low-frequency alternating current through the body e.g. muscle stimulation, ventricular fibrillation and burns, are covered in Chapter 3. Lightning and lightning injuries are discussed in Chapter 4. The effect of high-frequency (electrosurgical and diathermy) currents, and the type of injury encountered, are discussed in Chapter 5. Because current preferentially seeks tissues with the lowest resistivity, a knowledge of the resistivity values for tissues, organs and body fluids permits estimation of the current path; Chapter 6 provides such information. The same chapter presents information on the resistance of body segments and of a contacting conductor, showing the nonlinear nature of such contacts where tissues are boiled, charred and burned.

Because there are so many types of electrical accidents, it is hoped that the material contained in this handbook will allow the reader to understand the cause and the nature of specific electrical injury. The accident investigator should be able to identify the cause of an accident, whether due to carelessness or an act of God. The diagnostician and therapist need to know the effect of electric current and its pathway to select the appropriate therapy. It is hoped that all who have an interest in electrical accidents will find the information that they seek in this handbook.

Many have provided valuable information that has been included in this handbook, and the author hereby recognizes such assistance. I wish to thank Timothy Malloy, a patent attorney who taught me about the legal aspects of scientific information and what is expected of an expert witness in the tense atmosphere of the courtroom. Larry Conrad of Public Service Indiana Energy and Harold E. Amstutz, DVM of Purdue University are to be thanked for providing many papers and monographs on stray voltage. Dr. Amstutz also provided a wealth of literature on the environmental aspects of high-voltage power lines. Special thanks must go to Prof. Emeritus Theodore Bernstein of the EE Department of the University of Wisconsin for sharing the treasure chest of information that he has collected and so eloquently presented in papers and lectures; many of the latter I have attended and been entertained and informed by Ted's unique presentation style.

L.A. Geddes
Jan. 1, 1994

TABLE OF CONTENTS

Chapter 1
ELECTRICAL ACCIDENTS

INTRODUCTION

Generators that convert mechanical energy to electrical energy first appeared around the middle of the 19th century; electrical energy was first used to drive motors in factories and in locomotives. The outdoor gas lamps were replaced by electric arc lamps for street lighting in the late 1800's. However, it was the development of the carbon-filament electric lamp by Swan in the UK and Edison in the US in 1879 that brought electricity into the home. Edison in the US and Siemens in Germany were quick to sell electrical energy. Edison's direct current could not be transmitted efficiently very far from the generating station because of the need for large-diameter conductors. It was Tesla and Westinghouse who solved this problem by using alternating current, thereby eliminating the distance limitation with a transformer to step up the voltage for long-distance transmission with low current, requiring only small-diameter conductors. A step-down transformer at the user site recovered the electrical energy at any desired voltage.

HISTORICAL BACKGROUND

With the increasing availability of electrical energy from the mid 1850's, there arose the opportunity for accidents. Jex-Blake (1913) was the first to collect early accident reports and wrote: "I believe that no loss of human life from industrial currents of electricity occurred before 1879, though currents strong enough to have caused death were employed in lighting the operatic stage in Paris at the first performance of Meyerbeer's *Le Prophète* as long ago as 1849, and in lighthouses on and off the coast of England in 1857. In 1879 a stage carpenter was killed at Lyon by the alternating current of a Siemens dynamo that was giving a voltage of about 250 volts at the time. The man became insensible at once and died in twenty minutes; artificial respiration was not applied. The first death in this country (UK) took place at a theater in Aston, outside Birmingham, in 1880, where a bandsman short-circuited a powerful electric battery, became insensible, and died in forty minutes." As will be seen, modern electrical accidents have rather similar scenarios; Table 1.1 presents a recent summary of death statistics in the US (1987–89) due to electricity.

MICROSHOCK AND MACROSHOCK

The term "microshock" refers to cardiac arrhythmias produced by low-intensity current passing through the heart, usually via an intravascular or intracardiac catheter; microshock is discussed elsewhere in this book. The term macroshock is used to describe all other shocks. Electrocution refers to the loss of life due to electric current. Although there is no standard electrical accident, it is useful to provide an overview of the types of injury that can occur. The author prefers to distinguish between cases in which there has been obvious current flow through the body (or

Table 1.1: Deaths in the US due to electrical injury, 1987-1989*

SOURCE	1989	1988	1987
Domestic wiring and appliances	143	122	121
Generating plants, distribution stations, transmission lines	143	165	177
Industrial wiring, appliances and electrical machinery	61	75	64
Other and unspecified electric current	355	714	760
TOTAL	702	714	760

*From Accident Facts 1992

part thereof) and cases in which there was no obvious current flow through living tissue. Examples of the former are easily recognized. Examples in which there was injury with no current flow through the body would be eye or ear damage, or a burn resulting from being present when a sudden high-current discharge occurred, accompanied by a flash and an acoustic shock wave, as with an electric arc or lightning. Arc welders can sustain such flash injuries.

Injury can occur with a very low current flowing through the body (or a segment thereof) due to a startle reaction by the surprise encounter of a voltage source which produces a mild shock. If the subject were standing on a ladder when the shock was encountered, equilibrium can be lost and injury results from the fall, not due to the current flow. Physiological and pathological responses can result from electric current, depending on: 1) the current pathway, 2) type of current, 3) its duration of flow and 4) magnitude. Moderate power-line (50–60 Hz) current passing through the arms, legs and chest muscles can produce strong muscle contractions; pain often lasts long after current flow has stopped. Current flowing through the thorax can contract the respiratory muscles and impair or prevent continued breathing. A slightly higher power-line current passing through the chest can cause ventricular fibrillation, a condition in which all fibers of the main pumping chambers (ventricles) of the heart contract and relax randomly and cardiac output falls to zero. Irreversible brain damage starts to occur within three minutes. Therefore prompt cardiopulmonary resuscitation (CPR), followed by defibrillation, is essential for survival. Such current flowing through the respiratory center at the base of the brain can arrest breathing; again, prompt resuscitation is necessary for survival.

Higher current will produce heating and burns. The heating is proportional to the current squared and the duration of flow, as well as the area of contact. The concept of energy density factor, namely the current density squared multiplied by the duration of current flow (amps per sq. cm. squared multiplied by time), was introduced by Pearce et al. (1983) to provide a means for quantitating the severity of electrical burns due to radio-frequency current.

A more complicated scenario is associated with lightning injury, where there may be a burn as well as respiratory arrest and the loss of consciousness due to current flow through the brain.

JOULE HEATING, ARCS AND FLASHES

The thermal processes associated with electric current are Joule heating and arcs and flashes. Joule heating is the familiar temperature rise due to a current flowing through a conductor, examples of which are the electric toaster and stove. The temperature rise depends on the square of the current (I), the resistance (R) of the conductor and the duration of current flow (t in sec). The number of calories produced is $0.24I^2Rt$. One calorie is the energy required to raise one gram of water one degree Centigrade. The temperature rise (δT) in °C for a mass of m gram with a specific heat of S is $0.24I^2Rt/mS$, assuming no heat loss. Therefore, if current is injected into the body, without an arc occurring at the contact points, the intervening tissues will be heated. However the current flow between the contact points will not be uniform because of the differing conducting properties (resistivities) of the body tissues and fluids. The current density (amps per square centimeter) will be highest in the lowest-resistivity tissues. Moreover, the current density at the contact site will not be uniform.

In electrical accidents in which current flows through the body from skin-surface contacts, there is usually a thermal injury on the skin due to arcing at these sites. An arc represents the breakdown of an insulator (dielectric). All dielectrics, including air, have a limit to the voltage/cm that they can sustain without becoming ionized, resulting in an arc, i.e. an intense flow of ions and electrons, producing radiant energy in the form of light and heat. The temperature in an arc is typically 5,000 to 6,000°C. Therefore nearby flammable objects can be ignited. When the voltage between two conducting bodies in air exceeds about 75,000 volts/inch (pk), an arc is struck.

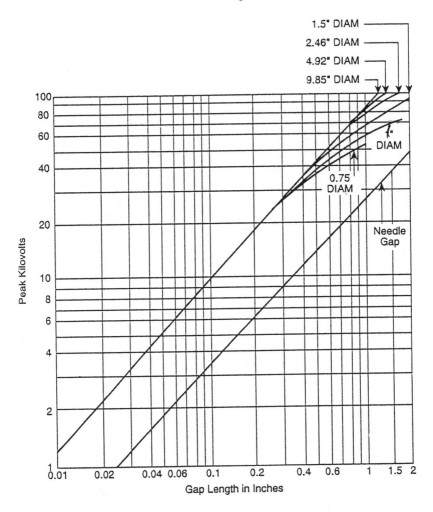

Figure 1.1 : The relationship between breakdown voltage and air gap length for air at 760 mmHg and 25°C (From ITT Reference Data for Radio Engineers 1968. Howard Sams Co. Inc., by permission)

This breakdown voltage depends on the barometric pressure, the temperature and the shape of the conductors. Arcing occurs at a lower voltage for pointed conductors. Figure 1.1 illustrates the relationship between breakdown voltage (peak) and air-gap length for 760 mmHg barometric pressure and 25°C for pointed (needle) conductors and smooth spheres. The breakdown voltage is approximately proportional to the pressure and inversely proportional to the absolute temperature in degrees Kelvin (ITT Handbook 1968).

When an arc is struck, the voltage between the conductors drops considerably. In fact it takes many times the voltage to strike an arc compared to that which sustains it. Bernstein (1993) stated that it requires about 50 volts/inch to sustain an arc with alternating current.

When an arc is struck there is an explosive liberation of energy which includes a shock wave. Privette (1993) reported that light-weight clothing was blown off a manikin at a distance of six inches from a one-foot arc carrying 8,000 amperes for 166 msec. He reported that the heat flux at six inches from such an arc is 171 calories per square cm-seconds. For 10 cycles of current (0.166 sec), the heat flux is 28 cal/cm^2. The radiant thermal energy emitted by such a pulse melted polyester underwear beneath protective clothing on a manikin. Note that these events were not associated with current flow through the manikin.

BURNS

The reciprocal nature of temperature and exposure time for a first-degree burn (skin reddening) was reported by Moritz and Henriques (1947) who used the pig, the skin of which is a good analog of human skin. Using a copper block, heated to different temperatures, they determined the relationship between temperature and exposure time for a first-degree burn, which is reddening of the skin, much like a mild sunburn. Figure 1.2 shows that the same first-degree burn can be produced by a high temperature presented for a short time as a low temperature for a longer time. The asymptote is the lowest temperature for a first-degree burn for an infinitely long exposure time which requires about 42°C. Curves for second and third-degree (full-thickness) skin burns were not obtained, but it is obvious that they would lie above the first-degree burn curve.

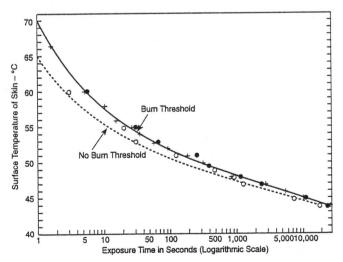

Figure 1.2: Temperature versus exposure time for burns.
(Redrawn from Moritz and Henriques 1947).

Another important aspect of a burn caused by a metal conductor that injects current to the skin is the nonuniform current density distribution thereunder; Nelson et al. (1975), Overmeyer et al. (1979), Caruso et al. (1979) and Pearce et al. (1983, 1986) all demonstrated this fact. Figure 1.3A illustrates the current density under a circular skin-surface electrode; note that the current density is highest under the perimeter of the electrode. Recall that the heating depends on the current density squared. Therefore the skin under the conductor perimeter will be hotter than under the center. Figure 1.3B is a thermogram obtained just after removing a circular electrode that delivered current into the skin. The high-temperature (white) ring identifies the perimeter of the electrode. Many accounts of electrical burns describe this characteristic pattern.

CURRENT FLOW AND CONTACT SITES

In all cases of electrical injury, the investigator seeks to provide a rational explanation for the event; contact sites provide useful information on the current path. The terms entry and exit sites are often used, being derived from the study of gunshot wounds where the entry wound is small and the exit wound is large; this terminology is not appropriate with electric current injury. If one contact site is small and the other is large in area, the injury will be more severe at the small-area contact site because it is the current density squared multiplied by the duration of current flow that determines the heating and hence the extent of injury.

In a particular case, the voltage is known, but the current and its duration are not. The current is equal to the source voltage divided by the impedance of the circuit, which consists of three parts: 1) that of the two contact sites, 2) that of the body segment through which the current flows and 3) the impedance of the voltage source, which in the case of power-line current, is low. Many investigations have been made of the impedance offered to power-line current applied at various sites on the body. It has been established that the circuit consisting of the contact sites and that of the body segment is nonlinear, i.e. the overall impedance decreases with increasing voltage, the reduction being the greatest for voltages up to about 200 volts. This reduction in impedance is due to breakdown of the dielectric (insulating) properties of the skin, as well as local heating. According to Lee (1977), from 200–500 volts the body-circuit resistance is relatively constant, amounting to 500–1,000 ohms. Detailed information on this relationship is found in Chapter 6.

When a high current flows for an appreciable time, the tissue is heated. Because living tissues are made up largely of electrolytes, it is useful to recognize that electrolytes have a negative coefficient of resistivity, amounting to about 2% decrease in resistivity for a 1°C increase in temperature. Therefore as the current flow is prolonged, the tissue temperature rises and its resistance decreases, resulting in a further increase in current and a further reduction in resistance etc. When the tissue fluids boil, there is an increase in resistance.

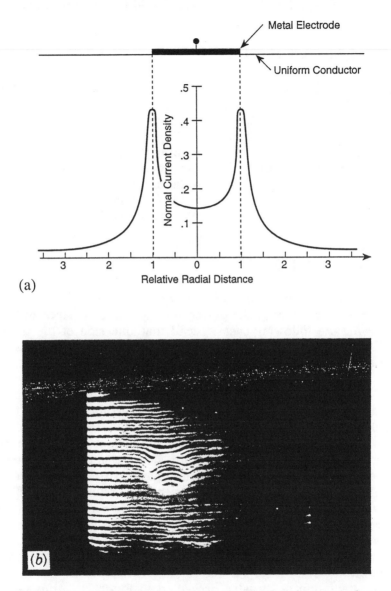

(a)

(b)

Figure 1.3: (A) Theoretical current density distribution under an electrode delivering current. Redrawn from Overmeyer et al., (1979). (B) Thermogram illustrating skin temperature under a circular disk electrode on the thigh after immediate removal of the electrode. (From Geddes and Baker, Principles of Applied Biomedical Instrumentation, 3rd edition, New York 1989. John Wiley & Sons. By permission.)

When an arc occurs, there may be a further increase in impedance if the current flow becomes intermittent. Therefore the events at a burn site can change rapidly; Chapter 6 provides additional information on this subject.

ELECTRICAL ACCIDENTS

Electrical accidents can occur in the home, workplace, hospital, indoors and outdoors. The source of the electricity can be the power line, devices connected to it or lightning. The types of electrical accident are many and can be bizarre. It has been said wisely that electricity and water do not mix. The presence of water not only provides a means for improving contact with the body, but also provides a ground-return path. Water issuing from a tap or hose provides an excellent ground. With these facts in mind it is useful to examine a few common accident scenarios. Additional explanations for these accidents can be found in the chapters that follow.

HOME ACCIDENTS

The home can be a hazardous environment because it has both a source of electrical energy (the power line), grounded devices and water. In reviewing the following cases, recall that one side of the domestic power line is grounded (see Chap 3). The letter G identifies cases investigated by the author.

Case 1G. An adult male came out of the shower and sat on the stainless steel counter surrounding the sink. He reached up and pulled the beaded metal chain on the light fixture to turn on the light. He received a shock, became unconscious immediately and fell to the floor. When the paramedics arrived about 10 minutes later, the victim was cyanotic (blue) and pulseless. He was given CPR (cardio-pulmonary resuscitation), after which the ECG showed coarse (low-frequency) ventricular fibrillation. Defibrillation was attempted but fibrillation occurred again and he was pronounced dead on arrival at the hospital. Obviously the beaded metal chain was in contact with the hot (ungrounded) side of the power line and the current flowed in the head-to-foot direction, producing ventricular fibrillation.

The invitation to disaster presented in Figure 1.4 illustrates the hazard when a grounded subject comes into contact with the 60-Hz power line via faulty insulation of the beaded pull chain that turns the light on. A similar scenario is popular in murder mysteries in which a subject in a bathtub is using a line-operated appliance, or one is thrown into the bathtub.

Figure 1.4 : Invitation to disaster; electricity and water do not mix.

Case 2G. A householder was standing on a metal stepladder in the base-ment starting to repair a light fixture. The switch was off and there was some water on the floor. When he touched one of the wires to the fixture, he received a strong shock and fell off the ladder, injuring his back. On investigation, it was found that the light switch was in the cold (ground) side, rather than the hot (ungrounded) side of the power circuit. His fall disconnected him from the power line and probably saved his life.

Case 3G. A householder was using a knife to dislodge a piece of toast that failed to emerge from an automatic toaster which was on a grounded metal surface. The knife touched the red-hot element and the toaster housing. A flash occurred, molten metal injured the householder's hand, and the fuse blew. The subject escaped with minor injury.

Case 4G. An adult female was using the telephone during a thunderstorm. There was a lightning strike nearby and the woman was thrown from her chair to the floor. She felt a shock to the head and complained of ringing in the ear with temporary hearing impairment.

It is important to recognize that lightning is characterized by a series of very short-duration pulses (see Chapter 4) and that an ohmic contact is not needed between the subject and the telephone. The metal parts in the headset and the hand and the head form the "plates" of a capacitor through which very short-duration pulses pass readily. Whether the muscular response was due to the current flow or the startle reaction due to the loud auditory sensation is not known. The subject returned to work on the following day and recovered completely.

There are reports of rodents chewing on the insulation of a two-conductor 120-volt, plastic insulated cable. When the insulation is gnawed through, the exposed conductors can pass current through the tissues or can become short circuited and an arc occurs, burning the rodent: such a spark can cause a fire. Rodents are not the only subjects that chew power-line cables; there are cases of children chewing extension cables. Severe oral lesions occur when the two conductors come in contact in the mouth.

INDUSTRIAL ACCIDENTS

Case 5G. At the completion of work at an outdoor construction site where cement had been poured, a worker was sent to clean out an electrically operated (230-volt) cement mixer with a water hose. His co-workers noticed that he had not returned and one was sent to check up on him. The worker was found several feet from the mixer, pulseless and not breathing with his feet and head in line with the opening of the mixer. The hose was found nearby, water still issuing therefrom. When the paramedic arrived, the victim could not be resuscitated and the ECG showed coarse ventricular fibrillation. The distance of the victim from the mixer is probably due to the strong muscular contraction due to the shock that he received from the metal parts of the mixer which was later found to be ungrounded and there was a high leakage current to the mixer frame.

Case 6G. Two workers were erecting an aluminum flagpole in the vicinity of an 11,000-volt transmission line. A gust of wind blew the pole into contact with the transmission line. Both workers were thrown to the ground and received hand burns. They were dazed but recovered. Obviously, the high-voltage and short duration of contact did not produce ventricular fibrillation.

A similar accident occurred when a crane came in contact with a high-voltage power line. When this occurred, the crane was raised to the potential of the high-voltage transmission line. The operator within the cab of the crane felt nothing because his environment was equipotential. However, a worker standing on the ground and touching the metal tracks of the crane received a shock.

Lee (1961) reported the following several types of electrical accident:

Case 13. A 60-year-old laborer, wearing rubber boots, was standing in a damp trench sawing through (an armored) cable believed to be dead. In fact, the cable was energized to 11,000 volts. There was a sudden flash and he became rigid and was thrown backward into the trench; the saw was destroyed. He felt very dazed and although able to climb into the ambulance, he did not really recover consciousness until he arrived in hospital a few minutes later. He was kept only a few hours and then sent home. As a result of the accident the backs of his hands were scorched, but not badly enough to require dressing. He was off work only for one day. As he was wearing rubber boots it is considered more likely that the path of the current was from right hand to left hand (his left hand was holding the grounded armored covering of the cable) than from hands to feet. He was not wearing gloves.

Lee reported a similar incident as follows:

Case 65. A joiner was using a wooden-handled hacksaw, held in the right hand, to cut through an 11,000-volt cable believed to be dead. He supported the cable with the toe of his left boot. There was a sudden explosion and a flash which vaporized the hacksaw blade. The joiner was quite definite that he did not receive a shock. The current had passed from the core of the cable and along the blade to the grounded armored casing of the cable. He was thrown out of the pit by the explosion, temporarily blinded by the flash, and was trembling all over and feeling very cold. He was given first-aid treatment (warmth and warm drinks) by his mate and taken to the nearest hospital where he was kept for three hours and then resumed work.

The two foregoing cases are similar, except that a shock was reported in Case 13 and no shock in Case 65. The fact that a high-voltage shock was received in Case 13 probably explains the lack of ventricular fibrillation (see Chap 3).

Lee (1961) described an interesting case in which the chest muscles were contracted strongly by arm-to-arm power-line current. He reported:

Case 19. A 21-year-old linesman, working outdoors in hobnailed boots, was standing on wet dewy soil, and grasped a copper wire in contact with a live conductor at 230 volts with the right hand. He could not let go and felt his right arm tightening, and then his chest. He later stated that he was losing consciousness when rescued and reported that his colleagues said he was going blue. They applied artificial respiration to him at once and he was taken to the medical unit of a nearby industrial firm (unfortunately the firm kept no record of this incident). He felt "shaky" for half an hour and rested for the remainder of the day although he felt all right.

For a week afterwards he felt a stiffness in his neck and chest as though he had unaccustomed exercise.

Case 81. A 54-year-old female assistant cook was cleaning the top of an electric cooker. She was wearing rubber shoes, she had her left hand on the top of the cooker and a wet cloth in her right hand. She felt a shock up her right arm and the hand contracted breaking the circuit immediately. She was quite well immediately after and lost no time from work. This was presumably a brief arm-to-arm shock. (Presumably due to leakage current).

Lee (1961) described the following burn associated with an electrical arc:

Case 58. A 31-year-old fitter's mate was working in a substation when a three-phase flashover occurred on the 11,000 volt circuit. He received burns to the back of his hands and front of his face, neck, and right wrist (these necessitated a period in hospital of about three or four weeks). The man remembers little of what happened, having only "islands" of memory until about three days after the accident. Artificial respiration was administered at the time of the accident by his colleagues.

HOSPITAL ACCIDENTS

In the environment of a patient experiencing medical treatment, three types of electrical hazard can be identified. One relates to sparks produced by static electricity or environmental instruments; a second results from the indirect contact with the power line energizing diagnostic, monitoring, therapeutic and assistive devices. The third type of hazard results from radio-frequency electrosurgical current, which is used to cut and coagulate living tissue. In some treatment areas, radio-frequency current is used to produce heat; the name for this technique is diathermy. These three types of hazard are variably present and depend on the particular circumstances, that is, whether the patient is in the operating room, coronary/intensive care unit, ward or specialized diagnostic or therapeutic area. Parker (1967), van der Mosel (1970), Feldtman and Derrick (1973), and Stanley (1974) have written extensively on this subject. It should be recognized that these environments have other hazards, such as infection and radiation of a variety of types; these however, will not be discussed. All accidents occurring in a hospital are investigated and followed by the filing of an incident report.

ELECTRIC SPARK HAZARD

Static electricity, one of the many causes of electric sparks, is the electricity of friction which is produced when different insulating materials are rubbed, the most familiar occurring when the hair is combed or when walking over a carpet on a dry day. Contact between the subject and a nearby metal object results in a spark which discharges the static charges separated by the energy of friction. Equally good static-electricity generators are clothes made of synthetic fibers. The potentials developed in such cases are in the tens of thousands of volts, and the spark discharge can be centimeters in length. A single spark can stimulate sensory receptors, nerve and muscle. In the latter case, if the discharge is applied directly to an exposed motor nerve or skeletal muscle, it can evoke a twitch. If such a discharge is applied to heart muscle it can evoke an extra systole. It is not known if such a discharge presented during the ventricular vulnerable period can evoke ventricular fibrillation when applied directly to the heart or the body surface.

Sparks from static electricity, motor starters and thermostats used to be serious hazards in the operating room when flammable anesthetics were used. Modern inhaled anesthetics are nonflammable.

FIRE

For a fire to occur, three ingredients must be present: 1) a source of flammable material, 2) oxygen and 3) a source of ignition (i.e. a spark or high temperature as produced by a laser beam). In the operating room all three ingredients can be present and circumstances can conspire to cause a fire or explosion; the following are a few selected examples.

Typically during anesthesia, 80–95% oxygen is used with the anesthetic agent applied to the airway via a plastic tracheal tube. Occasionally there may be a leak in an anesthetic circuit and oxygen accumulates around the patient who is covered with sterile paper or cloth drapes.

Case 7G. A defibrillation shock was delivered to two chest electrodes, one of which was not well applied and there was an arc under this electrode; the result was a flash that ignited the drapes, even though they were treated with a flame retardant.

Electrosurgical (high-frequency) current (see Chapter 5) is used to cut and coagulate tissue by the production of a small arc at the tip of an electrode in contact with (or just above) the tissue.

Case 8G. In an anesthetized subject with a plastic tracheal tube in place, a surgeon entered the trachea with an electrosurgical probe. The combination of a spark, oxygen and an flammable material (plastic tracheal tube) resulted in an explosion, melting the anesthetic-machine tubes and burning the patient.

The return path for electrosurgical current is via a large-area dispersive electrode on the subject. Occasionally this electrode is poorly applied or becomes dislodged, resulting in a small area of contact and a high current density at the skin, causing a burn. Sometimes the dispersive electrode is not applied or not plugged into the electrosurgical unit (sometimes called a Bovie), and the return path for the current is by any contact with a grounded object, such as the operating table, via ECG monitoring electrodes or an electronic rectal thermometer. At these sites, burns can occur due to what is called an alternate ground path. Many such burns have occurred (see Chapter 5).

BOWEL-GAS EXPLOSION

Explosion of gas in the gastrointestinal tract, ignited by an electro-surgical cutting electrode or by a laser scalpel, is not unknown. Levy (1954) reviewed much of the literature to that time and pointed out that digestive processes, bacterial fermentation, diffusion of gas (presumably inhaled flammable anesthetic gas) from the bloodstream, and swallowed gas are responsible for such accumulation. The composition of bowel gas is influenced by the amount of milk and legumes ingested (Table 1.2). From these sources, hydrogen and methane are produced; both are flammable. On the average diet, the bowel content for hydrogen is about 21%; for methane it is about 7%; and the carbon dioxide content is between 9–69%. On a milk diet, gut hydrogen increases to about 44%, and on a legume diet, methane also increases to about 44%.

Table 1.2: Bowel Gas

Diet	H_2 %	Methane %
Average	21	7
Milk	44	
Legume		44

Hussey and Pois (1970) reviewed the literature on bowel-gas explo-sions from the time of Levy's report (1954) to 1970, and described a case of bowel-gas explosion which occurred when the large bowel was opened with an electrosurgical cutting electrode. The bowel was ripped by the ex-plosion and required resection. Septicemia developed and the patient died.

Prevention of bowel-gas explosion merely requires attention to the hazard. For example, opening the bowel can be achieved with a scalpel to let the gas escape; thereafter the electrosurgical unit can be used. Some advocate purging the bowel with an inert gas such as nitrogen. Obviously dietary management of a patient is important prior to bowel surgery.

It would be difficult to find a better example of the disastrous combination of a combustible substance, oxygen and a source of ignition than that reported by Wegrzynowicz et al. (1992) who stated "A 35-year-old man presented for laser ablation of recurrent laryngeal papillomata. History and physical examination were unremarkable other than a 20-pack/yr smoking history, a bushy mustache, and a weight of 129 kg. Anesthesia was induced with thiopental and Fentanyl. Ventilation via mask with oxygen, nitrous oxide, and isoflurane was without difficulty. Vecuronium was given to provide relaxation. The surgeon inserted an adult Dedo laryngoscope, and jet ventilation was instituted with oxygen via a 13-G cannula inserted in the left light-carrier channel of the Dedo laryngoscope. A thumb-controlled valve and 50-psi oxygen powered the jet (were used).

"The patient's face and the perioral area were covered with soaking wet towels such that only the barrel of the Dedo laryngoscope was visible. Anesthesia was maintained with thiopental and Fentanyl during jet ventilation. There were no intraoperative problems except for brief periods of decreased hemoglobin oxygen saturation measured by pulse oximetry (S) during some episodes of apnea that were requested by the surgeon to eliminate movement of the vocal cords.

"Near the end of the surgical procedure, the surgeon suddenly yelled "fire," and bright blue and orange flames accompanied by a muffled roar were observed coming up through and around the laryngoscope. Jet ventilation was stopped; the towels were removed and the patient's blazing mustache was extinguished with the wet towels. The surgeon, who was in a great deal of pain, noted that the latex glove had been burned away from two of the fingertips of his right hand. The Dedo laryngoscope was removed, and bag-and-mask ventilation was commenced. Subsequent rigid bronchoscopy revealed no carbonaceous material in the trachea, and except for evidence of lasering, the glottis was normal. Muscle relaxation was reversed and the patient was awakened. Recovery from anesthesia was otherwise unremarkable.

"The patient suffered second-degree burns to his right upper lip and nasal rim. These were treated with 1% silver sulfadiazine (Silvadine) and healed over the next 2 weeks without further incident. The only other evidence of airway fire was burned nasal hair. The surgeon suffered second-degree burns to the right index and middle fingers that were severe enough to prevent him from operating for a week.

"This patient experienced an unusual complication. An errant laser strike on the surgeon's latex glove ignited the glove, producing hot volatile fuel that was entrained by the jet ventilator. Combustion of the vaporized latex accelerated dramatically in the oxygen-enriched atmosphere of the airway. The gaseous products of combustion escaped through either the patient's nose or mouth and in turn ignited his mustache, despite the "protective" wet drapes. It would be impossible to determine to what degree the patient's mustache contributed to his facial burns, or whether combustion alone under the drapes would have been adequate to cause the degree of injury sustained."

DIRECT CURRENT INJURY

Accidents have occurred due to direct current flowing through electrodes or devices attached to patients. The following cases provide examples.

Case 9G. A patient was prepared for surgery with two stimulating electrodes behind the knee and recording electrodes on the scalp to detect somatosensory evoked potentials (SSEPs) during surgery. The patient was draped and the equipment was connected to the electrodes. During surgery, the equipment functioned normally. However, somehow the stimulating electrode cable became disconnected from the stimulator. It was reconnected, but not to the stimulus output, but to the battery-test jacks which were connected to an internal 90-volt battery. The SSEPs were no longer recordable and at the end of the surgery, when the patient was being prepared for transport to the ward, a very large electrochemical "burn" was found at the site of the two stimulating electrodes.

Leming et al. (1971) reported the following case of an electro-chemical burn by low-voltage direct current.

"A forty-year old white female underwent surgery which lasted two hours. Before and during the administration of anesthesia, the patient was monitored with a cardioscope attached via four limb leads and a finger plethysmograph, the only electrical or electronic devices used throughout the procedure. Before the onset of anesthesia, the patient complained of a painful "pricking" sensation on the back of her left thumb, the finger to which the plethysmograph had been connected. The patient was reassured and put to sleep for the procedure. On the following day, in response to pain on the left thumb, examination revealed that the skin, which had been under the right leg electrode (the ECG ground lead) had multiple, discrete, punctate lesions not present before the operation. On the left thumb near the base of the finger nail, there was a dark gray, dry necrotic lesion with a small perforation near the center. On investigation, it was found that the 10-volt, direct-current supply within the photoplethysmograph resulted in a 10-volt potential difference between the right-leg (ground) ECG electrode and the plethysmograph case.

Leming et al. (1970) pointed out that "two ordinary flash-light batteries, for example, would be classified as harmless, but accidental or intentional long-term contact with these devices will result in burns," "especially if connected to devices in ohmic contact with a subject."

In conclusion, whether in the home, at an industrial site or in hospital, it is wise never to touch an electrically energized appliance with a metal housing while touching a water pipe. Although metal housings are supposed to be grounded, the ground connection may have been disconnected, but the appliance will still function normally and the hazard will not be noticed. In particular, the invitation to disaster shown in Figure 1.4 should never be allowed to occur.

POST-ELECTRICAL INJURY SEQUELAE

Following an electrical injury there may be long-lasting sequelae. Although it may be possible to estimate the amount tissue damaged in the current path, the victim's response may be disproportionally disabling. Some victims exhibit what is known as post-traumatic stress disorder (PSD). Following the accident, the victim is suddenly faced with an impairment that may threaten the ability to cope with the future, such as being able to return to the pre-injury job. This stress may produce symptoms of depression, manifested by a feeling of hopelessness, helplessness and a belief that no one understands. Anxiety, sleep disturbances and inability for prolonged concentration may be present. The degree to which these manifestations exist depend considerably on the pre-shock personality. Victims of electric shock, especially if the trauma is severe, need professional psychological support and rehabilitation.

WIRING CODES

In establishing the cause of an accident it is important to determine if the electrical wiring was in accordance with existing codes. In the US, such codes are to be found for situations, not under the control of electric utilities, in the National Electrical Code (NEC), published by the National Fire Protection Association (NFPA). The NEC is the bible for all electrical wiring and covers the workplace, home and health-care facilities as well as entertainment areas. There may be state, county or local regulations, which add requirements beyond the scope of NEC. Another code, the National Electrical Safety Code (NESC) is published by the Institute of Electrical and Electronics Engineers (IEEE). It covers systems owned by utility companies and deals primarily with high-voltage systems.

MEDICAL DEVICES

Legislation

A large number of diagnostic, therapeutic and assistive devices are electrically operated or controlled and if a malfunction occurs, a hazardous situation or accident can result. In the US, the safety and efficacy of all medical devices is under the jurisdiction of the Food and Drug Administration (FDA). In March 1976, legislation (HR-11124, Federal Register) was passed giving the FDA authority to control medical devices which were placed into three classes. Class I devices are nonhazardous by their nature and good manufacturing practice (GMP) controls their quality.

If placed in class II - performance standards - a device shall be required to meet an applicable standard on such date as is prescribed by the FDA, but not before one year after the date on which the standard is established. The major general controls will continue to apply to the device unless superseded by the standard. In effect, it means that enough data relative to safety and efficacy are available to accurately describe and control devices in this category. An important subtlety of class II is that safety and efficacy can be defined on the basis of published scientific data.

If placed in class III - premarket approval - and it is a new device, it may not be marketed until it meets premarket-approval requirements. If it is a device that was on the market before the date of enactment, a regulation must first be promulgated to require premarket approval and then the device manufacturer has until the later of 30 months after its classification or 90 days after the promulgation of the regulation to file an application.

There are special provisions for implantable devices and devices for supporting or sustaining life. The bill required panels to recommend that devices intended to be implanted in the human body which are on the market prior to the date of enactment of the bill - or which are substantially equivalent to such devices - be classified into class III - subject to premarket approval - unless they determine that such classification is not necessary to provide reasonable assurance of safety and effectiveness. It also requires that all devices including implantable devices not on the market prior to the date of enactment - and not substantially equivalent to devices on the market before such date - undergo premarket approval before they may enter the market.

Device Failures

Despite rigorous manufacturer's testing to assure the safety and efficacy of medical devices, accidents do occur associated with the proper or improper use by trained personnel. In association with proper use, a component failure can result in a malfunction that provides erroneous information or does harm to the operator or patient. When such events occur, the information manual that is provided with the device assumes special importance. In other words, does the manual advise of potential hazards, the need for special training or precautions? Often the labelling of the device is called into question. For example, does the device carry a label stating "to be operated by trained personnel only?"

Life-Support and Life-Sustaining Devices

Obviously, malfunction of life-support and life-sustaining devices constitute a hazard to a patient. For example, a ventilator that breathes for a patient may fail and an automatic alarm may not be sounded; if ventilator dependent, the patient may die unless help arrives. Likewise, if a defibrillator fails to deliver a shock to arrest ventricular fibrillation, the patient will die unless promptly resuscitated. Numerous other examples can be cited; the following illustrates a few.

An implanted cardiac pacemaker is a life-sustaining stimulator that causes contraction of the ventricles, the main pumping chambers of the heart. It is used when the ventricular rate is too low to provide sufficient cardiac output. Pacemakers operate under severe conditions and their present state of perfection is a tribute to the pacemaker industry which produces more than 100,000 per year in the US alone. Implanted cardiac pacemakers were introduced in the early 1960's (see historical background by Geddes 1990). With the first pacemakers, short battery life (about one year) and lead (catheter electrode) failure were the major difficulties. With introduction of the lithium battery, a 5 to 7-year pacemaker lifetime is now typical. The use of new alloys markedly reduced lead failure. However, there are still a few failure modes associated with implanted cardiac pacemakers. It should be noted that a pacemaker catheter electrode is subjected to both mechanical and electrochemical stresses. For example, with a heart rate of 72/minute, the number of heart beats per year is 37.8 million. With each beat, the catheter electrode is flexed, requiring the use of special alloys to withstand many years of operation. In addition to the large number of mechanical stresses, the catheter electrode operates in a hostile environment of blood, a reasonably good electrolyte. Two dissimilar metals in an electrolyte constitute a voltaic (galvanic) cell, the potential of which depends on the species of metals. Two dissimilar metals in contact with each other in an electrolyte constitutes a short-circuited voltaic cell which produces electrolysis and consumes the metals. In other fields, this phenomenon is known as galvanic corrosion. Thus in pacemaker leads, bimetal junctions must be protected from such corrosion by being well covered. Although pacemaker leads have evolved to a high degree of reliability, lead failures still occur, causing a loss of sensing or a loss of stimulation of cardiac muscle.

Because a pacemaker battery is inaccessible, it is not possible to test battery condition. Instead, a technique has been developed that allows determination of the battery condition. The technique employs a magnet placed on the skin over the pacemaker. The magnetic field operates a switch in the pacemaker and provides a pacing rate, known as the magnet rate which is measured very accurately at recommended intervals during the expected life of the pacemaker. The magnet rate decreases with time. Each manufacturer specifies a magnet rate for pacemaker replacement. Toward the end of the expected life of the pacemaker, magnet-rate checkups are typically at three-month intervals.

In the early days of pacemakers, their function could be disturbed by microwave ovens. This problem was solved by requiring shielding for such ovens and incorporating protective circuitry in pacemakers. However, during electrosurgery, pacemaker function can be disturbed. Modern programmed pacemakers usually revert to a fixed-rate or demand-mode operation if their programming is disturbed by external interference. Sometimes patients with implanted pacemakers require emergency transchest defibrillation. Such a life-saving procedure can damage a pacemaker, especially if it is of the monopolar type. A good review of pacemaker malfunction was presented by Hauser, 1994.

From the foregoing, it is clear that an implanted cardiac pacemaker is not only a sophisticated life-sustaining implanted stimulator, but it operates in a adverse environment under hostile conditions. When death or injury occurs in association with a pacemaker, it is important to discover whether the pacemaker was designed to function under the circumstances of the accident and if the bearer reported for frequent pacemaker checkups.

Conclusion

In all medical-device accidents, the first step is to obtain all of the facts. The incident report and interviews with those at the scene, including the victim (if possible) provide the most reliable information. If the accident was due to the failure of a medical device, a medical device report (MDR) is filed with the FDA[1]. The second step is to determine the reason for the accident. The third step is to determine how the circumstances were caused to be present. Finally it is necessary to determine if there was human negligence of if, despite all of the precautions taken, the accident would have happened.

REFERENCES

Bernstein, T. Personal communication 1993.

Caruso, P., J.A. Pearce, and D.P. DeWitt. Temperature and current density distributions at electrosurgical dispersive electrode sites. Proc. 7th N. Engl. Bioeng. Conf. 1979 pp. 373-374.

Feldtman, R.W. and Derrick, J.R. The hazardous hospital environment. Tex, Med. 1973, 69:63-67.

Geddes, L.A. Historical highlights in cardiac pacing. Eng. in Med. Biol., 1990, 9(2):12-18.

Hauser, R.G., Interference in modern pacemakers. Medtronic News 1994, 22(1):12–20.

Hussey, T.L. and Pois, A.J. Bowel-gas explosion. Amer. Journ. Surg. 1970, 120:103-105.

---

[1] FDA, 5600 Fisher's Lane, Rockville, MD 20857

ITT Reference Data for Radio Engineers, 5th ed. 1968 Howard W. Sams & Co., Inc. Indianapolis, Kansas City & New York.

Jex-Blake, A.J. Death by electric currents and lightning. Brit. Med. Journ. 1913, Mar 1: 425-430.

Lee, W.R. Lightning injuries and death. In Golde, R.H. Lightning. Vol 2. London 1977 Academic Press.

Lee, W.R. A clinical study of electrical accidents. Brit. Journ. Ind. Med. 1961, 18:260-269.

Leming, M.N., Jacob,, R.G. and Howland, W.S. Low-voltage direct current plethysmograph burns. Med Res Eng 1971, Oct-Nov; 1921.

Leming, M.N., Ray, C. and Howland, W.S. Low-voltage, direct-current burns. JAMA 1970, 30:1681-1685.

Levy, E.I. Explosions during lower bowel electrosurgery; method of prevention. Amer. J. Surg. 1954, 88:754.

Moritz, A.R. and Henriques, F.C. The relative importance of time and surface temperature in the causation of skin burns. Am. J. Pathol. 1947, 23:605-720.

National Electrical Code, National Fire Protection Association (NFPA), Battery-mark Park, PO Box 9146, Quincy, MA, 02269-9959.

Nelson, J.A., Gagliano, L.C. and Clements, D.D. Current distribution at electro-surgical ground sites. Proc. 28th Ann. Conf. Eng. Med. Biol. 1975, 25:12.

Overmeyer, K., Pearce, J.A. and DeWitt, D.P. Measurements of temperature distribution at electrosurgical dispersive electrode sites. Trans. ASME 1979, 101:66-72.

Parker, B. Electrical testing for safety of the operating room and intensive care unit. Amer. Coll. Surgeons Bull. 1967, 54:187-189.

Pearce, J.A., Geddes, L.A., VanVleet, J., Foster, K. and Allen, J. Skin burns from electrosurgical electrodes. Med. Instrum., 1983, 17:225-231.

Pearce, J.A. Electrosurgery. 1986, Chapman & Hall, London, 270 pp.

Privette, A. Burn injury resulting from electrical flashovers. In Electrical Injury. June 11-12 1993, Chicago IL. Sponsored by the Univ. of Chicago, Amer. Burn Assn. and the IEEE.

Stanley, P.E. Safety in the electromedical equipment system. Natl. Safety News 1974 (Nov.) 71-75; (Dec.) 80-89.

van der Mosel, H.A. Electrical safety and our hospitals. Med. Instr. 1970, 4:2-5.

Wegrzynowicz, E., Jensen, N.F., Pearson, K. et al. Airway fire during jet ventrila-tion for laser excision of vocal cord papilloma. Anesthesiology 1992, 76:468-469.

Chapter 2
STIMULATION OF EXCITABLE TISSUE AND SENSORY STIMULATORS

INTRODUCTION

To appreciate the many and varied responses of excitable tissue to electrical stimuli,it is useful to start with a presentation of the basic facts about electrical stimulation embodied in the strength-duration curve, which is a plot of the lowest (threshold) current (I) for excitation with a single rectangular pulse of current of duration (d), versus the duration of the pulse. The effect of multiple pulses and other waveforms will be dealt with subsequently.

STRENGTH-DURATION CURVE

Weiss (1901) and Lapicque (1909, 1920) were the first to enunciate the fundamental law of excitation, which states that the shorter the duration (d) of the current pulse, the higher the current (I) required to stimulate. Lapicque introduced the terms rheobase (b) and chronaxie (c) and an empirically derived mathematical expression for the strength-duration curve shown in Figure 2.1A. The rheobasic current (b) is the threshold for an infinitely long-duration pulse. The chronaxie (c) is a tissue-dependent constant which Lapicque defined as the duration for which the threshold current is twice rheobase, i.e. 2b; these quantities are identified in Figure 2.1A.

Blair (1932) formalized the law of excitation by deriving a mathematical equation based on the known resistive and capacitive properties of cell membranes. The Blair expression for the strength-duration curve is:

$$I = b/(1-e^{-d/\tau})$$

where I is the peak current for a rectangular pulse of duration d; b is the rheobase and τ is the membrane time constant for the particular tissue. By choosing $I/b = 1.59$ on the strength-duration curve, the duration becomes the membrane time constant (τ), which is specific for each type of tissue. Table 2.1 presents typical values for τ. Figure 2.1C presents strength-duration curves for cardiac muscle, sensory receptors and motor nerves. Note that the shapes of the curves are similar and the rise with decreasing duration is determined by the membrane time constant.

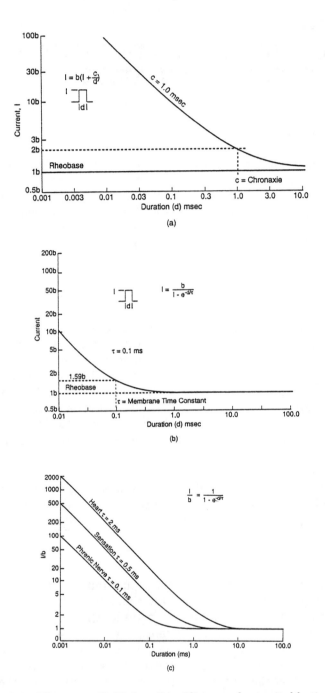

*Figure 2.1. The strength (I) duration (d) curve for excitable tissue.
In A is shown the Lapicque, and B shows the Blair model. In C are shown
curves for different tissues.*

Table 2.1: Membrane Time Constants[*]

Tissue type	Time constant (msec)	Temp °C	Investigator and year
Nerve			
Human (motor)	0.015[a]	Body	Ritchie (1944)
Cat (A fiber)	0.022	29.5	Li and Bak (1976)
Cat (C fiber)	3.1	29.5	Li and Bak (1976)
Cat (C fibers)	0.80[b]	37	Koslow et al. (1973)
Cat (C fibers)	0.85[b]	27.5	Koslow et al. (1973)
Pain fibers			
Human (skin receptors)	1.0	Body	Notermans (1966)
Human (glans penis)	0.28	Body	Adrian (1919)
Sensory receptors			
Human (skin)	0.33	Body	Ayers et al. (1987)
Human (sensory nerve)	0.12–0.36	Body	Adrian (1919)
Skeletal muscle			
Human (denervated)	2[b]	Body	Ritchie (1919)
Cardiac muscle			
Dog (ventricle)	2.2[c]	37	Geddes et al. (1987)
Dog (ventricle)	13.2[c]	25	Geddes et al. (1985)
Dog (ventricle)	2.14 ($\pm$0.97)	37	Pearce et al. (1982)
Sheep (purkinje)	4.3	37	Dominguez and Fozzard (1970)
Sheep (purkinje)	3.75	37	Fozzard and Schoenberg (1972)
Turtle (ventricle)	7.3 ($\pm$2.22)	Room	Pearce et al. (1982)

[a] Recent studies provide larger values.
[b] Estimate.
[c] Same animal.
[*] From Geddes, L.A. and Baker, L.E. Principles of Applied Biomedical Instrumentation, 3rd edition. New York 1989 John Wiley. 961 pp.

MECHANISM OF STIMULATION

All excitable cells are enveloped by a selective, ion-permeable membrane which, at rest, results in a high concentration of potassium (K^+) ions in the cell and a high concentration of sodium (Na^+) ions outside the cell as shown in Figure 2.2A. The net result is a transmembrane potential, the outside being positive with respect to the interior. At rest, the transmembrane potential is called the resting membrane potential (RMP), as shown in Figure 2.2B, and is typically 70–90 mV.

For excitation to occur, it is only necessary to reduce (ΔV) the RMP by about one third by removal of positive charge from the cell membrane (hypopolarization) by a cathodal pulse of current (I). During the current pulse, the RMP reaches the threshold potential (TP) and the membrane permeability increases suddenly, accompanied by a rapid influx of Na^+ ions and efflux of K^+ ions, resulting in a cyclic excursion in membrane potential, which rises steeply (excitation) and returns to the RMP more slowly (recovery). Both Na^+ and K^+ ions are subsequently restored by metabolism. In Figure 2.1C are shown strength-duration curves for heart muscle (τ=2 msec), sensation (τ= 0.5 msec) and motor (phrenic) nerve (τ= 0.1 msec), normalized for current by division by the rheobase (b).

The membrane time constant (τ) is determined from the strength-duration curve. By putting d = τ, the ratio for I/b = 1.59. In other words, by identifying the point on a strength-duration curve where I/b = 1.59, the corresponding point on the duration axis is the membrane time constant.

IMPLICATIONS OF THE STRENGTH-DURATION CURVE

Figure 2.2B depicts the intimate process of stimulation resulting from injecting a negative (cathodal) pulse of current I, that reduces (hypopolarizes) the resting membrane potential (RMP) to the threshold potential (TP), which causes the cell to produce a propagated action potential. For a stimulating current to flow, there must be a second electrode and, although not shown in Figure 2.2A, this electrode was large and distant from the active electrode that injected the current I. Such a technique is called monopolar stimulation. Because it is current density (mA/cm^2) that stimulates, excitation of the cell occurred under the small-area electrode that injected the current I.

If the small, current-injecting cathodal electrode is moved away from the excitable tissue and placed on the skin, as shown in Figure 2.3A, the current will spread and the current density at the excitable tissue will be less for the same current. Therefore to achieve stimulation, the current must be increased to attain a current density sufficient to reduce the resting membrane potential to the threshold potential. Figure 2.3B presents strength-duration curves for stimulating the heart with a direct myocardial electrode and an electrode on the pericardium, the thin membrane that envelops the heart. Note that the curves are similar, but the rheobasic value (b_e) for the extrapericardial electrode is higher than that (b_m) for the myocardial electrode.

From the foregoing, it is possible to generalize that the rheobasic current depends on the distance to the excitable tissue from the stimulating electrode. The current for stimulation with shorter duration pulses will always be above the rheobasic current, the rise in threshold with decreasing duration depends on the membrane time constant (τ) of the excitable tissue.

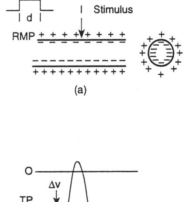

(a)

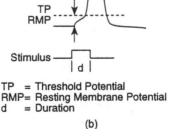

TP = Threshold Potential
RMP= Resting Membrane Potential
d   = Duration

(b)

Figure 2. 2. In A is shown the charge on a typical cell membrane. In B is shown the membrane potential being decreased by a rectangular pulse of current I of duration d, resulting in the genesis of an action potential.

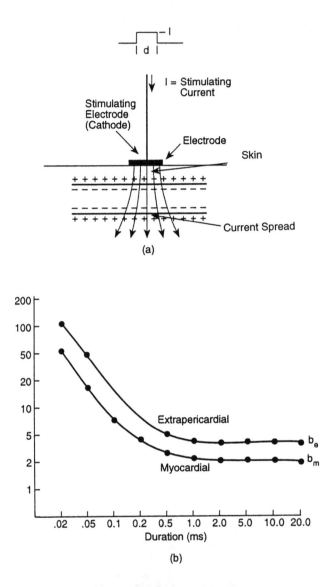

Figure 2.3. Stimulation of excitable tissue with a cathodal current pulse applied to an electrode that is removed from the tissue (A). In B are shown strength-duration curves for stimulating heart muscle with an electrode on the heart (myocardial) and with an electrode more distant on the pericardium (extrapericardial), showing that the current required to stimulate is higher when the electrode is more distant.

ANODAL STIMULATION

It is well known that positive-polarity pulses will stimulate excitable tissue. Recall that it requires two electrodes to complete a circuit to obtain current flow. Figure 2.4A illustrates the situation where an anodal (positive) current pulse is applied to excitable tissue. Under the positive current-injecting electrode the outside of the cell membrane becomes more positive (hyperpolarized). However, the other side of the cell faces the other (cathode) electrode and becomes hypopolarized, i.e. less positive and if the current density is sufficient, excitation will occur at this site. Figure 2.4B shows strength-duration curves for anodal and cathodal current pulses used to produce sensation with a skin-surface electrode. Note that the anodal curve is above the cathodal curve.

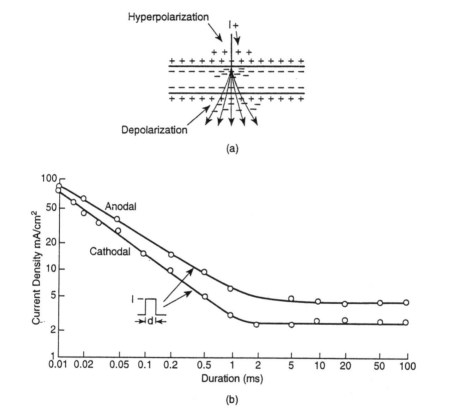

Figure 2.4. In A is shown stimulation with a positive (anodal) pulse of current which results in hyperpolarization of the part of the excitable cell nearest to the electrode and hypopolarization of the other side of the cell facing the other (distant) cathode electrode. Excitation occurs in the region that is sufficiently hypopolarized. In B are shown current-density (J)-duration (d) curves for sensation with single anodal (positive) and cathodal (negative) current.

REFRACTORY PERIOD

Before discussing the response to a train of pulses or alternating current, it is necessary to establish the concept of refractoriness. Figure 2.5A illustrates a typical action potential and during excitation (upstroke), and for a little more than half of the recovery (downstroke) of the action potential, the cell cannot be excited by a stimulus, no matter how intense; this period is called the absolute refractory period (ARP). Following the ARP, the membrane potential has recovered enough and the cell will respond to a suprathreshold stimulus. Slightly later it will respond to a subthreshold stimulus; still later normal (100%) excitability returns. Figure 2.5B illustrates the threshold for stimulation throughout the duration of the action potential.

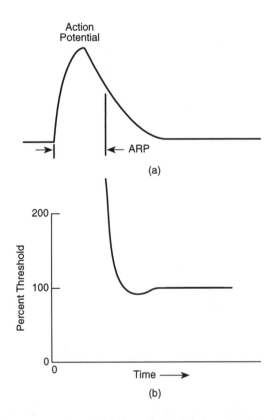

Figure 2. 5. Action potential (A) and excitability during the action potential (B). During the upstroke and early downstroke of the action potential, the cell is completely inexcitable; this period is called the absolutely refractory period (ARP). After the ARP, the cell regains its excitability, first to a suprathreshold stimulus, then to a subthreshold stimulus followed by normal 100% excitability.

RESPONSE TO REPETITIVE STIMULI

Whether or not a cell responds to each pulse in a train of stimuli depends on the interval between the pulses, i.e. the period (1/frequency) and the refractory period of the excitable cell. For example, if the refractory period is 0.1 sec, the maximum responding rate is 10/sec. Figure 2.6A illustrates a cell responding with a stimulus having a frequency such that its period is equal to the refractory period; note that there is one response per stimulus. Figure 2.6B shows the response when the frequency is higher; observe that there is not a response for each stimulus because stimuli occurred during the refractory period.

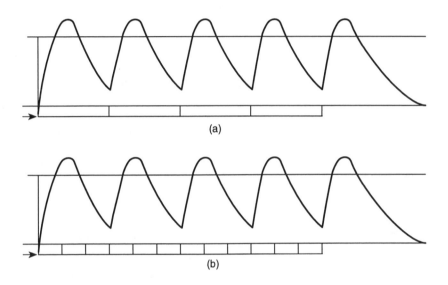

Figure 2.6. *Action potentials evoked by a train of stimuli whose period (1/frequency) is equal to the refractory period (A), thereby producing a 1/1 response. In B are shown the action potentials produced by a train of stimuli whose period is one third of the refractory period. Note that only every third stimulus evokes an action potential.*

RESPONSE TO SINUSOIDAL CURRENT

With the knowledge that both anodal and cathodal current can stimulate and that the refractory period of a cell will allow repetitive excitation, it is easy to see that sinusoidal alternating current can stimulate excitable tissue. Whether each cycle stimulates depends on the relationship between the period (T = 1/frequency, f) of the alternating current and the refractory period of the tissue. To a first approximation, if the equivalent duration (d) of sinusoidal alternating current is 1/f, it is useful to examine the strength-duration curves for cathodal, anodal and sinusoidal alternating current using the same electrodes on the same subject; Figure 2.7 presents such a comparison for sensation.

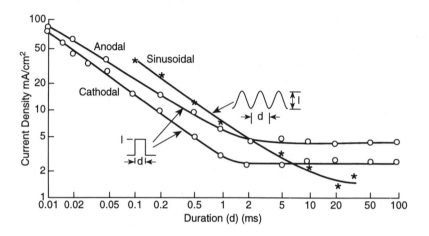

Figure 2.7. Strength-duration curves for sensation with single cathodal and anodal rectangular current pulses and with sinusoidal current, the duration being expressed as the period (1/frequency) and current expressed as the peak-to-peak value.

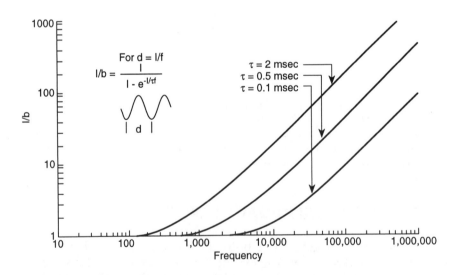

Figure 2.8. Normalized strength (I)-frequency (f) curves for sinusoidal current for tissues with different membrane time constants. Current is expressed as the threshold current (I) divided by the rheobasic (b) low-frequency threshold.

STRENGTH-FREQUENCY CURVES

From the foregoing evidence it is reasonable to equate the duration of a sine wave to 1/frequency; therefore the strength-frequency curve becomes

$$I = b/(1-e^{-1/\tau f})$$

Figure 2.8 presents calculated strength-frequency curves for tissues with different membrane time constants (τ). Note that in all cases, the current required for stimulation is higher with increasing frequency.

It is useful to examine experimentally obtained data which presents the magnitude of the alternating current required for sensation for increasing sinusoidal frequency. Dalziel (1956) reported the first studies on the threshold for perception of sinusoidal current with copper-wire electrodes applied to the hand; Figure 2.9A presents his results, showing the range of sensitivity expressed as percentile. Figure 2.9B presents similar data for neck-abdomen and transchest electrodes. The former are used in impedance cardiography and the latter are for recording respiration by impedance change.

When discussing skin-sensation threshold, it is useful to recall that it is current density (mA/cm^2) that stimulates and that the current threshold will be different for large and small-area electrodes. In addition the different skin sites will have different thresholds. On the right of Figure 2.9B are shown perception current-density thresholds for the neck-abdomen (NA) and transchest (TC) electrodes. Note that in all cases, the current density required for sensation increases with increasing frequency. A similar relationship for stimulating rat motor nerve was reported by Lacourse et al. (1985) who found that the strength-frequency curve was continuous up to 1MHz.

SKELETAL MUSCLE STIMULATION

The skeletal muscle response to a single stimulus is called a twitch; the response to a train of stimuli is called a tetanic contraction or tetanus. As the frequency of the stimuli is increased from say 1/sec, the twitches become closer and fuse into a tetanic (sustained) contraction; Figure 2.10 shows such a sequence. Note 1) that in this muscle, above about 10/sec the individual twitches have fused and 2) above about 30/sec, the force of the tetanic contraction does not increase, 3) the force developed by a tetanic contraction is much greater then that for a twitch and 4) the duration of a tetanic contraction persists as long as the stimulus train is delivered, providing muscle metabolism can be sustained. Tetanic stimuli applied directly to a motor nerve produce a very strong muscle contraction. The response to power-line current (50–60 Hz) is a tetanic contraction of skeletal muscle, the contraction persisting as long as the current flows.

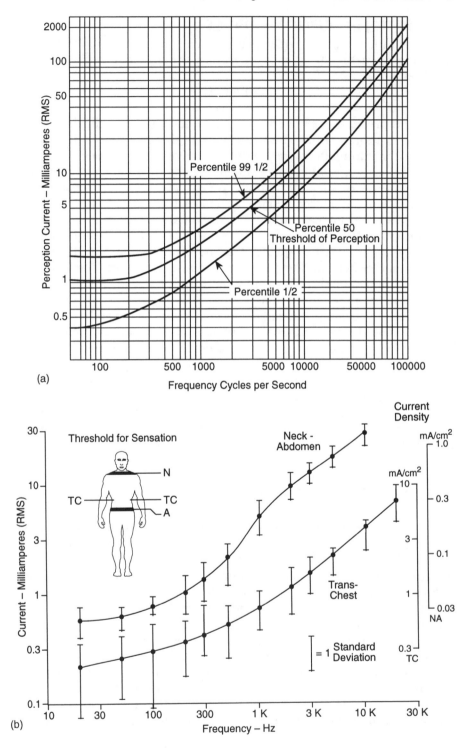

(a) Frequency Cycles per Second

(b) Frequency – Hz

Figure 2.9 (previous page). Sensation-frequency curves for elect-rodes on the hand (A), transchest (TC): electrodes and neck-abdomen (NA) electrodes (B). (A is from Dalziel, IEEE Trans. BioMed. Eng. 1956, PGME5:48- and B is from Geddes and Baker, L.E. Journ. Assoc. Adv. Med. Instr. 1971,5:13–18, both by permission)

Dalziel (1956) introduced the term "let-go current" to describe the ability of a subject to release himself from tetanic muscle contraction produced by alternating current. Using a hand-held #6 copper wire electrode, paired with a brass plate under the other hand or foot, he measured the maximum current that the subject could sustain and still release his grasp. Figure 2.11 shows the let-go current versus frequency for men, women and children. Observe that the lowest let-go current was found in the 10–200 Hz range. For 60-Hz current, Dalziel stated that although there was considerable variability among subjects, the average values for let go current (rms) for men was 15.87 mA and for women was 10.5mA.

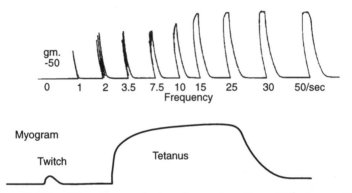

Figure 2.10. The force of muscle contraction with increasing stimulus frequency from a single stimulus to a train of 50/sec (A). In B is shown the response to a single stimulus (twitch) and the response to a train of stimuli (tetanus).

CARDIAC MUSCLE STIMULATION

The response of cardiac muscle to electric current pulses merits special attention. Pulses delivered at a rate of about 1–3/sec are used in cardiac pacing, each pulse producing a heart beat. Pulses delivered at a slightly higher rate can prove catastrophic by producing ventricular fibrillation, a condition in which all of the muscle fibers of the ventricles contract and relax randomly and blood pumping ceases. Ventricular fibrillation in humans does not cease spontaneously, and irreversible brain damage ensues if cardiopulmonary resuscitation is not applied within a few minutes, followed by defibrillation.

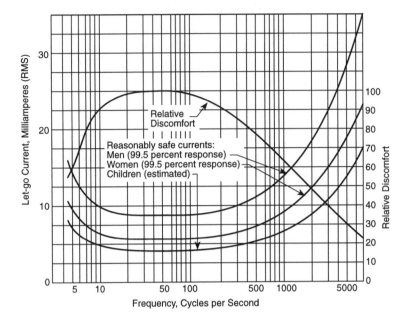

Figure 2.11. Reasonably safe let-go current versus frequency for men, women and children (estimated), along with a relative discomfort curve (From Dalziel, IEEE Trans.BioMed.Eng. 1956,PGME 5:44-62, by permission).

To understand how cardiac muscle responds to a current pulse, it is useful to recognize that the heart consists of two double pumps, i.e. two atria and two ventricles, the latter being the two main pumping chambers, as shown schematically in Figure 2.12. The atria pump blood into the ventricles, the right ventricle pumps venous blood into the lungs and the left ventricle pumps oxygen-rich blood into the arterial system. The atria are low-pressure, muscular pumps that fill the ventricles, the principal pumping chambers. The excitability characteristics of the atria and ventricles are described by their membrane time constants (τ).

To illustrate the effect of electric current on the heart, it is necessary to examine the action potential of the ventricles which underlies genesis of the R and T waves of the electrocardiogram (ECG), as shown in Figure 2.13. The action potential of the ventricles is shown in Figure 2.13B and the electrocardiogram that results from all of the ventricular action potentials is shown in Figure 2.13C. In Figure 2.13A is shown a series of strength-interval curves which identify the threshold current required for excitation at different times during the cardiac action potential (Fig 2.13B) for single rectangular pulses 0.1, 0.5 and 5 msec duration. Note that the refractory period, (from the onset of the action potential to the point where the curves rise steeply), is slightly different for the different duration pulses. Observe also that at the end of the action potential (100%), the current required for shorter pulse durations is greater, being evidence of the strength-duration curve.

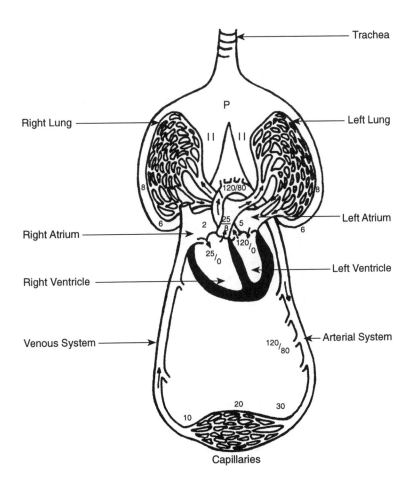

Figure 2.12. The circulatory system.
The numbers refer to the pressures in mmHg.

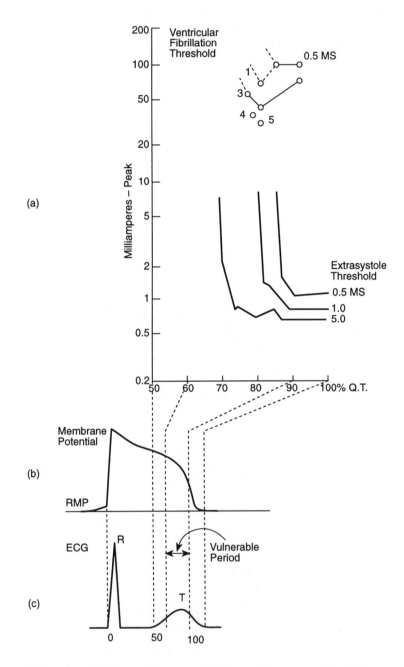

Figure 2.13.   Strength-interval curves (A) for evoking a ventricular contraction and ventricular fibrillation with a single stimulus applied during the vulnerable period. In B is shown a typical ventricular action potential and in C is shown the ECG that is derived from the ventricular action potential.

VULNERABLE PERIOD

Cardiac muscle can be excited by a single pulse of current during recovery after about 60% of the duration of the action potential (Fig 2.13B) and with such a stimulus, a contraction occurs. However, if a single strong pulse of current is delivered between about 60 to 80% of the duration of the ventricular action potential, fibrillation occurs; this period of hyperexcitability is known as the ventricular vulnerable period. The intensity of single pulses of different durations needed to induce fibrillation are shown in the upper portion of Figure 2.13A.

Figure 2.14A shows the ECG and blood pressure; at the arrow a single suprathreshold pulse of current was delivered which precipitated ventricular fibrillation, evidenced by the dramatic change the ECG and cessation of pulsatile pumping. When the ventricles are observed during fibrillation, a series of waves of rapidly moving contraction and relaxation can be seen, often the ventricles appear to shimmer. Figure 2.14B illustrates the location and extent of the vulnerable period with respect to the T wave of the ECG.

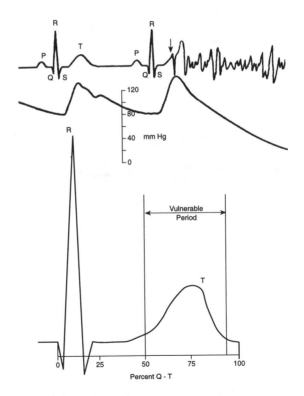

Figure 2.14. Ventricular fibrillation produced by delivery of a single stimulus during the vulnerable period (A) and, the extent of the vulnerable period in relation to the QRS and T waves of the ECG (B).

EFFECT OF POWER-LINE CURRENT ON THE HEART: VENTRICULAR FIBRILLATION

Figure 2.15A shows a sketch of a series of ventricular action potentials produced by a short train of low-intensity, power-line current cycles. Note that there are fewer action potentials than cycles of the power-line current because of the long duration of the ventricular refractory period. Figure 2.15B shows the ECG and blood pressure and at the arrow, 60 Hz alternating current was applied to the surface of the ventricles. After about 5 seconds, the ventricles fibrillated, cardiac pumping ceased and the blood pressure fell to virtually zero.

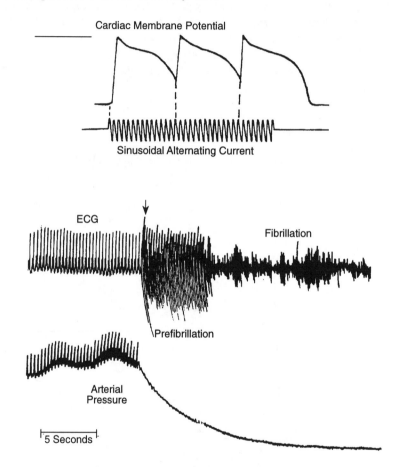

Figure 2.15. Cardiac action potentials produced by low-intensity power-line current (A). The ECG and blood pressure (B) before applying 60 Hz power-line current to the left ventricle. Note that after about 5 sec of tachycardia, the ventricles fibrillated and all pumping ceased.

The lowest current for precipitating ventricular fibrillation is via an electrode directly in contact with the ventricle. Such a microshock circumstance can arise if current is conducted by a saline-filled catheter or an electrode in the left or right ventricle. Geddes et al. (1971) conducted a dog study in which the ventricles were fibrillated by applying alternating current to intracardiac catheters; Figure 2.16 illustrates the experiments and the results. Observe that the lowest current for fibrillation was obtained with frequencies below 100 Hz. Above this frequency, more current was required. The low-frequency (60-Hz) current ranged from about 50 to about 325μA rms, the range probably representing the proximity of the electrode to the myocardium.

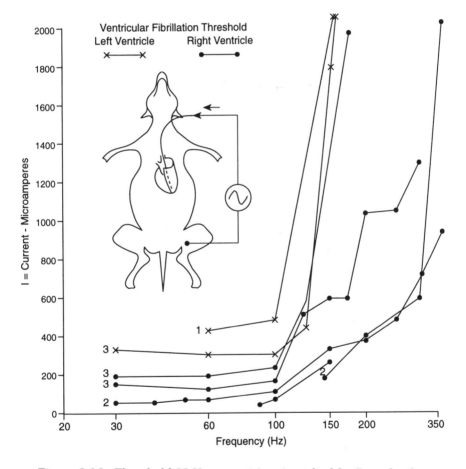

Figure 2.16. Threshold 60-Hz current (rms) applied for 5 sec for the precipitation of ventricular fibrillation with catheters in the left and right ventricles (From Geddes and Baker. Med. Instr. 1971, 5(1):13-18 by permission).

A considerable amount of research has been devoted to determining the lowest current required to precipitate ventricular fibrillation with catheter electrodes. An important study by Roy et al. (1977) is shown in Figure 2.17. Using electrodes of different areas in contact with the canine ventricles, they determined the threshold current for fibrillation with different durations of current flow; Figure 2.17 illustrates that the current required is lowest for about 5 sec or more of current flow and for the smallest area electrodes. Many other studies have been conducted to establish the threshold 60-Hz current for ventricular fibrillation with different electrode sites; Table 2.2 presents a summary.

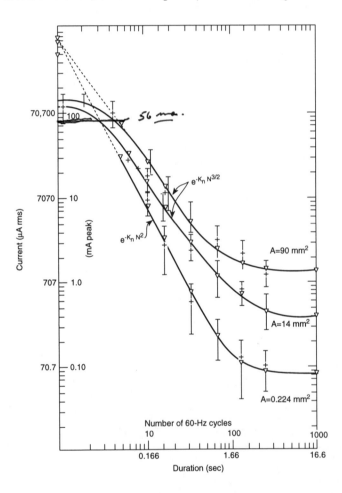

Figure 2.17.   Threshold 60-Hz current required for ventricular fibrillation with direct-heart electrodes and for current applied for different durations. Vertical bars indicate standard deviation for (+) experimental average values and (∇)-calculated values. (Redrawn from Roy, O.Z. et al., 1977).

Table 2.2: 60-Hz Ventricular Fibrillation Thresholds for Dogs

Exposure Time	Current (μA rms)		Electrode Type and Size	Reference
	RV	**LV**		
Increase to fibrillation	140–205 (min. 60)	110–162 (min. 75)	#7 catheter #7 Pt electrode	Weinberg and Artley (1962)
2 sec	258 (av) ($\pm$200 variability)		#7 catheter electrode	Whalen et al. (1964)
Not given	20		Bipolar catheter electrode	Staewen at al. (1969)
5 sec	60–430	300–430	#7F catheter	Geddes and Baker (1971)
10 sec	428[a]	485[a]	Catheter	Lee and Scott (1973)
			Catheter electrode	Starmer and Whalen (1973)
2 sec	2990		5.06 (cm^2)	" "
	1290		1.19 (cm^2)	" "
	340		0.85 (cm^2)	" "
15 sec	64–1190		Catheter	Roy et al. (1976)
>4 sec	1130		0.9 cm^2	Roy et al. (1977)
	296		0.14 cm^2	
	64		0.00224 cm^2	

[a] assumed to be rms current.

With cardiac catheterization being so common, it is important to recognize that a saline-filled catheter in a ventricle constitutes an electrical communication with the myocardium. Monsees and McQuarrie (1971) pointed out that if 60-Hz leakage current is carried by such a catheter, ventricular fibrillation could be induced. Using a value of 70 ohm-cm for saline, a 100 cm 8F catheter would have a resistance of 0.44 megohms. Assuming that 50 μA rms current is needed to induce fibrillation, the minimum 60 Hz voltage is $50 \times 10^{-6} \times 0.44 \times 10^6 = 22$ volts rms. This calculation assumes that the catheter tip is in firm contact with the ventricular myocardium. Higher current would be needed if the catheter tip were distant from the myocardium.

It should be noted that a catheter in either the beating left or right ventricle can come in contact with the endocardium and produce an ectopic beat. By such mechanical stimulation, it can produce ventricular tachycardia which can progress to fibrillation in an excitable heart. Therefore in a case of catheter-induced, ventricular fibrillation, it is necessary to establish if the underlying cause was due to electrical current or repeated mechanical stimulation due to the flailing catheter in the beating heart.

SUMMARY OF STIMULATION

The fundamental law of excitation is embodied in the strength-duration curve which states that for a single current pulse, the shorter the pulse duration, the higher the current needed for stimulation. The more distant the point of current application, the more the current needed for stimulation. With a train of pulses, there will be a one-to-one response if the period (1/frequency) of the stimuli in the train is longer than the refractory period of the tissue being stimulated. The refractory period is the earliest time after an effective stimulus when a second stimulus can evoke a response.

A single stimulus delivered to a motor nerve or muscle will produce a muscle twitch. Repetitive stimuli delivered to a motor nerve or skeletal muscle will cause a tetanic (sustained) contraction for as long as the stimulus train is applied. In the case of heart muscle, especially the ventricles, a train of pulses will induce ventricular fibrillation, resulting in a loss of pulsatile pumping and death, if prompt resuscitation is not applied. In addition, a single stimulus, delivered during the vulnerable period of the ventricles, can precipitate ventricular fibrillation. The foregoing provides information on the type of response that can be expected with single and multiple pulses of current applied to excitable tissue. The chapters that follow will use this information to explain the type of responses encountered when various types of current are injected by body-surface contact.

DIRECT-CURRENT APPLICATIONS

There are situations in which direct current is accidentally or intentionally applied to the body. Electrochemical lesions and ventricular fibrillation have been reported with the accidental application of direct current. Transdermal drug delivery is facilitated by low-intensity direct current; the technique is called iontophoresis. The following paragraphs provide examples of the response to such direct-current applications.

ELECTROCHEMICAL BURNS

This chapter has dealt with stimulation using current pulses and alternating current, demonstrating that excitation requires the sudden removal of a critical charge from the membrane of an excitable cell. With the presentation of direct current of adequate intensity, an excitable tissue will respond at the instant of current application. Despite continued flow of an unvarying direct current, there will be no excitation. However there will be a response when the current flow is suddenly terminated. Although there will be no characteristic excitation during the flow of a constant current, electrolysis will occur at the electrode-tissue interfaces. At the anode (+) oxygen and chlorine gas and hydroxyl ions are liberated; while at the cathode (−), hydrogen gas and sodium will be liberated. In this process, the current fluctuates as the gas bubbles leave the electrode-subject interface. The fluctuating current can cause stimulation. If the current is high enough and flows for long enough, nasty lesions can occur at the electrode sites; these are called electrochemical burns. Selected electrochemical burn accidents have been presented in Chapter I. The following describes the case in which direct current is intentionally introduced into the body.

In the early days of electricity, Prevost and Batelli (1899) examined the response of a variety of animals in which the direct current was applied to electrodes in the mouth (+) and rectum (−). They wrote:

"The dogs died by paralysis of the heart with relatively low tensions (50 to 70 volts), despite that respiration continued for a few minutes. The ventricles presented fibrillary tremulations of which we have spoken in a previous communication. The auricles [atria] continued to beat. It is consequently useless in this circumstance to practice artificial respiration.

"With higher voltages that we had available (550 volts) the heart was arrested by a single shock (closure and breaking), respiration is suspended for a few seconds but it restarts very feebly and superficially.

"To provoke fibrillary tremulations of the ventricles with a constant current, it is necessary to use a tension [voltage] of less than (50–70), the electrodes being placed in the mouth, in the rectum and on the thigh. Then only a current of 10 volts suffices with a.c. Then with a.c. the duration of contact must be at a minimum of one second, with the continuous current a single shock (knowing the time necessary to open and close the current must be around 0.1 sec.) suffices to obtain this result." (Author's translation).

ELECTROLYTIC LESIONS

The use of direct current to produce lesions of predetermined size dates from the last century when it was desired to determine the function of various areas in the brain by observing changes due to the destruction of specific brain cells. This literature was reviewed by Geddes (1972). From this review it is possible to estimate the size of an electrolytic lesion produced by direct current injected by a needle electrode. Although most of the lesions studied were produced in animal brains, the results in other soft tissues are not expected to be very different. Most of the studies employed needle electrodes about 0.5 to 1 mm diam. insulated down to a small tip. Without regard to polarity (which is of some importance in the studies that were performed carefully), the range of current multiplied by time varies between 20 to about 90 mA-sec; these parameters produced spherical lesions having diameters of about 1 to 1½ mm. There is ample evidence that with high currents (i.e., greater than about 5 mA) and long times (in the range of minutes), the lesion diameter is not proportional to the product of milliamperes and seconds; it is somewhat smaller. Lesions with flat plate electrodes will be larger than the extent of the plate and the severity of the injury will be the greatest under the perimeter of the electrode where the current density is highest.

TRANSDERMAL DRUG TRANSPORT

When a soluble substance is placed on the skin, it will be absorbed slowly. The rate of transport is proportional to the concentration difference across the skin. Initially, the concentration below the skin is zero and with the passage of time the concentration rises and the transdermal concentration difference decreases, thereby reducing the rate of drug transport; this is Fick's law of diffusion. If the capillary blood flow removes the subdermal drug, it will continue to be absorbed. In the late 1800s it was discovered that the application of direct current to an electrode placed over a drug-coated skin site facilitated the transdermal transport of a drug. Licht (1967) reviewed the fascinating history of this technique. Iontophoresis and electro-osmosis are two processes that enhance the transdermal transport of drugs by direct current; these processes will now be described.

Iontophoresis

Iontophoresis employs a direct current to transport a drug through the skin. Because an electrical field is also present, the transport of water is facilitated, this process being known as electro-osmosis. Iontophoresis transports charged drug molecules; electro-osmosis facilitates the transport of neutrally charged molecules by virtue of enhanced water transport.

The method of applying iontophoresis is shown in Figure 2.18, in which the drug to be transported is contained in a skin-surface (active) electrode (A), within which is an inert metal electrode that establishes electrolytic contact with the drug solution. The other, inert (dispersive) electrode (B) is much larger in area and at a convenient distant site. Between the two electrodes is connected a constant direct-current source. The polarity is selected on the basis of the polarity of the charge on the drug molecule. For example, if the drug carries a positive charge, the active electrode is made positive. The opposite polarity is used if the charge on the drug molecule is negative. The pH of the solution in the active electrode is important for efficient drug transport.

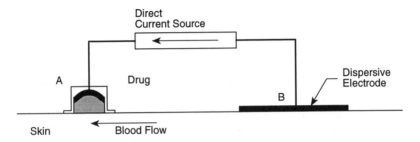

Figure 2.18. Principle of iontophoresis in which a constant direct current is used to transport drug molecules in an electrode chamber (A) into the skin. The return path for the current is via a large-area, dispersive electrode (B).

To avoid producing an unpleasant sensation, the current density (mA/cm^2) under the active electrode is kept low and the current is increased slowly to its selected value to avoid sensation. Likewise, when the treatment is to be terminated, the current is slowly reduced to zero before the electrodes are removed. If the current is turned on or off suddenly, the subject may feel a sensation like a pin prick or stronger if this current density is high.

The amount of drug transported through the skin by iontophoresis depends on the current, its duration of flow and the concentration of the drug in the active electrode chamber, as well as the pH. Treatment times usually last from 10 to 30 minutes. Currents up to 10mA have been used and the treatment is quantitated in terms of charge, i.e. the product of milliamperes and minutes. The current density is usually much less than 1mA/cm^2. High current density will produce an unpleasant pricking sensation.

Many different metals have been used for the active electrode which can be either positive or negative, depending on the charge on the drug molecule. Silver, tin, zinc, stainless steel, copper and carbon-loaded silicone rubber have all been used. Because direct current is used in iontophoresis, a wide variety of electrolytic products can be produced at the active electrode. For example if silver is used and made positive, it is chlorided; if it is made negative, silver ions are transported into the skin. Edelberg (1963) deposited silver into the skin to create a (black) skin-surface electrode which remained there for some time. The presence of silver in the skin is called argyria. Copper ions have been intentionally deposited in the skin for their germicidal effect. Such heavy metal ions combine with proteins. If the current is intense enough and lasts long enough, pain and skin sloughing can occur. For iontophoresis, it is wise to use an inert electrode material, such as stainless steel, platinum, gold or carbon and keep the current below the sensation threshold.

Iontophoresis has been applied to the eye, ear, nose, gums, genital areas and skin. Tyle (1986, 1988) reviewed many of the clinical uses for iontophoresis. When applied to the skin, alcohol is used to clean the site for the active electrode. Some of the uses for iontophoresis are the relief of localized cutaneous pain, creation of local anesthesia to obtain a shave biopsy or to cauterize superficial blood vessels. The production of local anesthesia is popular in dermatology. For example Maloney et al. (1992) compared the topical anesthetic effect of iontophoretically injected lidocaine and lidocaine combined with epinephrine in 64 patients. It was found that the mean duration of anesthesia was 12.1 minutes for lidocaine alone and 86.9 minutes for the combination of lidocaine and epinephrine. This 7-fold enhancement is due to the epinephrine which is a vasoconstrictor that prevents the rapid removal of the lidocaine by reducing the superficial capillary blood flow.

Electro-Osmosis

Electro-osmosis is defined as the movement of a fluid by an electric field. It is believed that electro-osmosis plays a role in the transport of drug molecules with a neutral charge. Gangarosa et al. (1980) reported that the transport of water through the skin is enhanced under the anode (positive) electrode.

Electrochemical Burn Hazard

The possibility for an electrochemical lesion exists with iontophoresis and electro-osmosis. Tingling or a prickling sensation under the active or dispersive electrode is an indication that the current is too high and should be reduced, the effectiveness of the treatment being restored by prolonging the time of the reduced current flow and/or increasing the concentration of the drug in the active electrode.

SUMMARY OF CURRENT LEVELS

In the preceeding paragraphs, the various types of responses to alternating and direct current have been presented. Dalziel (1956) presented data that summarizes the direct and alternating current levels for different effects; Table 2.3 presents such data.

Table 2.3: Effects of Direct and 60-Hz Alternating Current[*]

Effect	Direct Current (mA)		60-Hz Current (mA rms)	
	Men	Women	Men	Women
No sensation on hand	1	0.6	0.4	0.3
Slight tingling. Perception threshold	5.2	3.5	1.1	0.7
Shock- not painful and muscular control not lost	9	6	1.8	1.2
Painful shock - painful but muscular control not lost	62	41	9	6
Painful shock - let-go threshold	76	51	16.0	10.5
Painful and severe shock - muscular contractions, breathing difficult	90	60	23	15

[*] From Dalziel, IEEE Trans. Bio. Med. Eng. 1956, 5:44–62.

SENSORY AND STUNNING STIMULATORS

Several types of impulse stimulators are used to provide a moderately strong stimulus to alter behavior. Among such devices are the electric-fence controller, which keeps livestock within or without an enclosure, the cattle prod, the training collar that is used for obedience training, the Stun Gun and the Taser; the latter two are used to stop a criminal act without inflicting bodily injury. The characteristics of such devices, which produce pulses of high voltage and low current, will now be described.

ELECTRIC-FENCE CONTROLLER

An electric fence is designed to keep animals in or out of an enclosure by avoidance training, by delivering a shock when the animal contacts the fence and ground. Terms such as fence charger or fencer are used to describe such stimulators. Usually a single bare wire, mounted at nose level using insulators on fenceposts, provides the active electrode; an indifferent rod electrode driven into the ground provides the return path for the current.

Before the advent of standards for fence controllers, it was customary to place a low-wattage 120-volt incandescent lamp in series with the hot side of the domestic power line and the fence wire. The lamp was meant to provide a current limit when the animal contacted the fence. However, the cold resistance of the lamp filament is quite low and the desired current limit, based on the illumination resistance, is not achieved. Therefore, use of this technique is extremely dangerous because the stimulus is 60-Hz alternating current which flows for the duration of contact with the electrified fence.

Modern electric-fence controllers produce high-voltage, current-limited shocks at a rate of about one per second. An Underwriters Laboratory specification (#69) sets the performance standards; typically the maximum time on is 0.1 sec and the maximum current is limited to 4mA for an on period of 0.1 sec.

Electric-fence controllers are available in many sizes, being capable of electrifying up to 50 miles of fence. Smaller units are designed to keep animals out of domestic gardens. Those used for livestock and predator containment can produce a voltage pulse of 8,000 volts. There are types that contain batteries that are recharged by solar cells (Parker McCrory Co., Kansas City, MO 64108).

CATTLE PROD

A cattle prod or stock prod is a hand-held, flashlight-sized, battery-operated stimulator used by livestock handlers to keep stock moving in a desired direction; Figure 2.19 illustrates two such devices: PULSAR and POWER-MITE, provided by Hot-Shot, Savage MN, 55378. On the right of Figure 2.19 are oscillograms showing the output waveforms of three units. Note that the output voltages are high and the pulse durations are short. Typically the frequencies range from 50 to 500/sec. The output impedance is high and the current depends, in part, on the impedance of the load across the output terminals (1,2). Cattle prods are very effective in keeping animals moving. The shock is highly localized to the region of the electrodes (1, 2) and provides a strong sensory stimulus with some muscle contractions if the electrodes are in the region of a motor nerve.

TRAINING COLLAR

Animals learn quickly to avoid an unpleasant experience. If an animal performs an undesired act that is followed immediately by an unpleasant sensation, the animal will refrain from performing the same act. Such avoidance training is easily achieved by delivering an electric shock immediately after the undesired act. Figure 2.20 illustrates how this principle is applied using a collar that carries a radio receiver (A) that controls a stimulator connected to the two blunt conical electrodes (1, 2) which project from the collar. The radio receiver is commanded to deliver an attention-getting stimulus by a hand-held transmitter (B), the size of a flashlight. When a button on the transmitter is pressed, the transmitter sends a signal, via its protruding antenna, to the collar receiver which delivers a series of high-voltage pulses to the collar electrodes. Each stimulus is 300–400 μsec and the frequency is 200–300/sec. Only a short burst of stimuli is needed to get the animal's attention. The training collar shown in Figure 2.20 is extremely effective in controlling barking in dogs. The shock does not produce any noticeable cardiac or respiratory distress in tests conducted to date by the author.

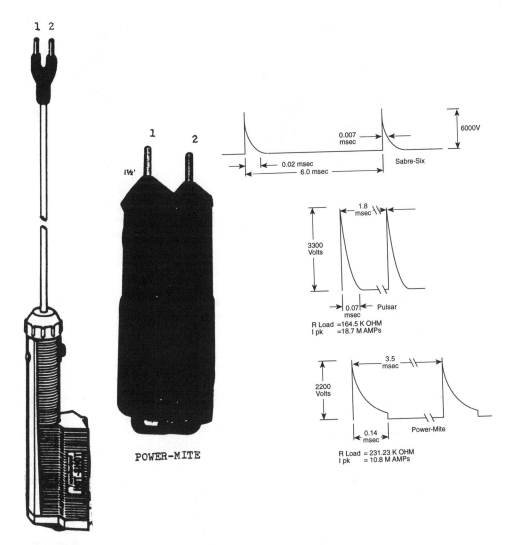

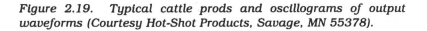

Figure 2.19.   Typical cattle prods and oscillograms of output waveforms (Courtesy Hot-Shot Products, Savage, MN 55378).

Figure 2.20. Training collar (A) and transmitter (B) that is used to command delivery of stimuli to the blunt rod electrodes (1, 2) in the collar (Courtesy Tritronics, Inc., Tucson, Arizona).

STUN GUN

The Stun Gun (Figure 2.21) is a hand-held device that delivers a localized stimulus that is strong enough to stun a person, thereby facilitating capture. When the contact probes (2 inches apart) are placed close to the skin, preferably over a muscle, a strong surprise sensation is produced with minor muscle contractions. Only about one quarter second of application is needed and the subject usually loses balance. The presence of clothing does not diminish the effectiveness because the high voltage pulses penetrate clothing easily. A 5 to 7-second application causes a victim to fall to the ground and leaves him weakened and dazed. Two red spots, resembling insect bites, may appear on the skin and disappear in a few days. However a 1–2 second current application is usually enough to cause a subject to fall. It is not advisable to apply the shock to the head. The NOVA XR5000 (Nova Technologies Austin TX, 78758) Stun Gun delivers a high peak voltage (90,000) train (17-22/sec) of short-duration pulses. According to tests conducted by Bernstein (1976), the XR 5000 unit delivers a train of damped sinusoidal stimuli (70 kHz). The energy in each pulse depends on the subject resistance. In static tests into a resistive load, the energy ranges from 52 mJ for a 200-ohm load to 119 mJ for a 1,000-ohm load.

Although stun guns are effective, there have been reports that some subjects under the influence of drugs become enraged and even more violent when shocked. However, such cases are by no means common.

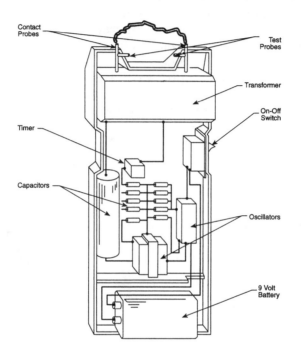

Figure 2.21. The Stun Gun
(Courtesy Nova Technologies, Austin, TX, 78758).

TASER

The Taser (Tasertron, Newport Beach, CA 92658), is a hand-held, nonlethal stimulator that ejects two dart electrodes from a cassette. Two slender wires (15') follow the electrodes which engage a victim's clothing. The dart separation is one foot for each five feet of range. Figure 2.22 illustrates use of the Taser and a sketch of the device.

The output of the Taser is 50,000 volts with a maximum power of 5 watts. The frequency of the short-duration pulses ranges from 8 to 22/sec; the energy per pulse is approximately 0.5 joule, the peak current being 100 μA. It is not necessary for the dart electrodes (1.4 gm) to penetrate the skin because the high voltage will cause the current to be carried by the tiny arc that forms. About 3 sec of current flow will incapacitate a subject who falls to the ground and is dazed and confused for a few minutes. However, when downed, the subject is able to obey commands.

Many safety tests have been conducted with human volunteers and animals. The shock delivered by the TASER encompasses a fairly wide region that stimulates both sensory nerves and muscles. When the subject is struck and the TASER is activated, there is a complete loss of balance and the subject falls to the ground because all voluntary muscle control is temporarily lost. For this reason, the response is not affected by prior use of drugs. The subject is dazed and confused for minutes depending on the activation time of the shock.

The TASER cannot be purchased by the general public because a rifle primer is used to propel the dart electrodes. Therefore the TASER is classified as a Title 1 firearm and has the same restrictions as a pistol. It has proven to be an effective, nonlethal firearm in many capture scenarios.

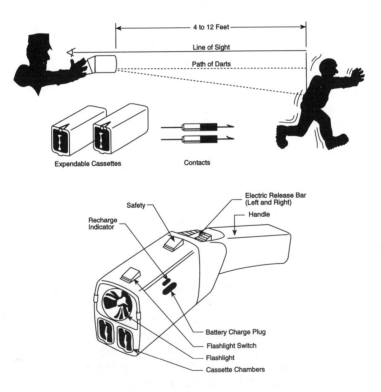

Figure 2.22. The Taser
(Courtesy Tasertron, Newport Beach, CA 92658)

IGNITION SHOCKS

High-voltage, low-current pulses are used to ignite the fuel-air mixture in internal combustion engines. Typically, the current is suddenly broken in the low-voltage primary of an induction coil with a secondary having many turns. The high rate of change of current (di/dt) produces a high-voltage pulse. The current pulse is a short-duration, damped sine wave, not unlike the pulses produced by the stunning devices just described. In fact, many of such devices employ automobile or motorcycle ignition coils. The response to such shocks have just been described.

MAGNETIC (EDDY-CURRENT) STIMULATION

Current can be caused to flow within the body without being contacted by conductors. By using a high-intensity, time-varying magnetic field, it is possible to induce a pulse with enough current density to effect stimulation. The first to do so was Arsène d'Arsonval, who in 1896 placed the head of a subject in a large coil carrying 30 amps of 42-Hz current. d'Arsonval wrote: "There occur, when one plunges the head into the coil, phosphenes and vertigo, and in some persons, syncope ... The alternating magnetic field modifies the form of muscular contractions, and produces in living beings, other effects that are easy to demonstrate and of which I am pursuing the study of at this time." (Author's translation).

Phosphenes are bright spots in the visual field, due either to stimulation of the retina or visual (occipital) cortex. The fact that muscle contractions were identified indicates that the motor cortex may have been stimulated. Numerous investigations followed some time after d'Arsonval's report and the history of magnetic, i.e. eddy-current stimulation, was reported by Geddes (1989).

Mechanism of Stimulation

The method by which a magnetic field can cause stimulation resides in Faraday's law of induction which states that a time-varying magnetic field will induce a voltage in a conductor lying within the magnetic field. Note also that if a conductor is moved within a static magnetic field, a voltage will be also be induced. Faraday's law underlies the operation of all transformers and generators of electric current. Figure 2.23 illustrates schematically the method by which current is induced in living tissue in the vicinity of a coil (L) carrying a time-varying current (i) produced by the voltage source E. The current (i^1) induced in the tissue is proportional to the rate of change of the current (di/dt) in the coil, i.e., i^1 = k(di/dt), where k is a quantity that depends on the coil geometry and the proximity of the tissue to the coil. The strength (B) of a magnetic field is measured in Gauss (G) or Tesla (T); one Tesla is 10,000 Gauss. The strength-duration curve applies to magnetic stimulation and the analog of current intensity is dB/dt, the rate of change of the magnetic field (B).

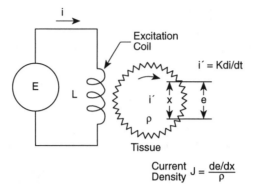

Figure 2.23. A time-varying current (i) flowing in a coil (L) will induce an eddy current (i) in tissue of resistivity ρ. The induced current is proportional to the rate of change of current (di/dt) in the coil (L).

Applications

There are a few applications of magnetic (eddy-current) stimulation, sometimes called electrodeless stimulation. These applications arose because the skin sensation with magnetic stimulation is much less than with skin-surface stimulating electrodes. Although many research studies have been carried out in which stimulation of the heart and peripheral motor nerves have been described, the only clinical use at present is stimulation of the motor cortex of the brain to evoke muscular contractions on the opposite side of the body. These clinical studies have been reviewed by Chokroverty (1989).

Electrical Hazards

There has been considerable interest in the possible effects of static and time-varying magnetic fields on living organisms. Some of the research in this area has been reviewed by Wadas (1991) and there remains much speculation in this area. In a practical situation when magnetic fields are applied to a human subject, there are three aspects that have been addressed, 1) sensory stimulation, 2) motor nerve and brain cortex stimulation and 3) cardiac stimulation. With motor-cortex stimulation, there is the danger of producing convulsion if the frequency of the pulses is above about 5/sec. The effect of much higher frequencies has been discussed in the chapter dealing with electronarcosis. Single pulses produce only muscular twitches, sensations, phosphenes or a taste sensation. In magnetic resonance imaging (MRI), considerable research has been conducted to examine the safety of the procedure; this research will now be described.

MAGNETIC RESONANCE IMAGING (MRI)

With magnetic resonance imaging (MRI), both static and time-varying magnetic fields, plus a radiofrequency field, are used to construct a tomographic image of internal structures. The subject is placed in a large cylindrical enclosure (Fig 2.24) in which there is a strong static magnetic field (1.5 – 2 T) in the head-foot direction. In research investigations, weaker and stronger (4T) fields are used. To form the image, time-varying currents are applied to three sets of (gradient) coils. Resonant radiofrequency magnetic fields excite protons selected by the gradients. The decay of the resonating protons in the tissues creates the raw signal for the tomogram. Fourier transforms of the raw signals over sample number and over the incremented gradient strengths produces the (2 or 3 dimensional) image. The resonant radiofrequency for protons is 42.57 MHz/T.

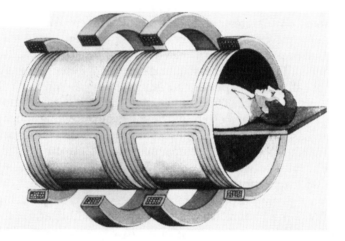

Figure 2.24. Magnetic resonance imaging equipment.
(Courtesy General Electric Medical Systems, Milwaukee WI)

Safety of MRI

The issue of the safety of MRI has been examined carefully, particularly with a view to the possibility of increasing the current in the gradient coils to enable image acquisition in a shorter time and to investigate new imaging techniques. Schaefer (1988) discussed many of the safety aspects and provided data on the body-temperature rise due to the radiofrequency field. This subject is discussed in detail in Chapter 5. Suffice it to say that the body-temperature rise limit set by the FDA is 1°C. The more recent literature on stimulation with magnetic fields will now be reviewed.

Static Field

Schenck et al. (1992) investigated the effect of the strength of the static magnetic field used in MRI. Two field strengths (1.5 and 4T) were used, the former (1.5T) being below the present 2T limit set by the FDA. The report stated: "Eleven healthy male volunteers were examined by an experienced industrial physician and a battery of standard laboratory and clinical examinations (blood and urine studies, EEG, EKG and psychometrics) were carried out. The volunteers were then intermittently exposed to the 4T field during imaging and spectroscopy experimentation. No episodes of acute distress were associated with such exposure and immediately postexposure no significant changes in blood pressure or core temperature (measured at the tympanic membrane) were noted. During approximately 12 months, 90 separate studies, amounting to 100 hours of exposure, were recorded. Total individual exposure ranged from less than one hour up to 35 hours. After 12 months the physical examinations and the other studies were all repeated. No significant abnormalities were detected in any of the volunteers on any of the examinations either before or after the period of high field exposure.

"Although no deleterious health effects were detected, several volunteers reported certain sensory effects at 4T." Pursuing the sensory responses, Schenck et al. submitted a questionnaire to "9 volunteers with 4T experience and to 24 others who had only 1.5T experience. They were asked whether they had experienced any of 11 sensations or related effects during field exposure. Three categories - nausea, vertigo and metallic taste - were reported more often by the 4T group to a statistically significant degree ($p < 0.05$ by Fisher's exact method). For 7 categories - balance, headache, muscle spasms, numbness, tinnitus, vomiting and hiccuping - the difference between 1.5T and 4T respondents was not significant at the $p = 0.05$ level. Another sensation - magnetophosphenes or flashing lights occurring during rapid eye motion - was noted only during special conditions (darkened room) at 4T. By gradually withdrawing the volunteers from the field this effect was found to vanish once the field at the eyes had fallen to about 2T. We attribute this effect to motion-induced retinal excitation associated with the extremely low voltage threshold of the photoreceptors." The report continued with "The sensory effects were associated with motion in the field, were mild or moderate in severity and would not normally interfere with patient studies.

"We suggest that the sensations of vertigo and nausea result from magnetohydrodynamic forces on the endolymphatic fluid in the semicircular canals induced by motion of the head through the magnetic field. No field-induced auditory sensations were reported. This supports the attribution of the effect to the semicircular canals rather than to direct excitation of hair cells." In concluding, the report stated: "These results suggest that there is a substantial safety margin for human exposure to conventional imaging fields of 1.5 to 2T, that there are nonharmful sensory effects associated with motion in static magnetic fields, that these effects are near or somewhat above their threshold values at 4T, and that some individuals experience them at even lower field strengths."

As stated earlier in this chapter, current is induced in a moving tissue in a static magnetic field. The author has experienced weak phosphenes when moving the head very vigorously or the eyes rapidly in a static 1.5T magnetic field, (with no gradient-coil current); this experience is consistent with that provided by Schenck.

Time-Varying Magnetic Fields

Using a standard Signa (GE) body coil with no static magnetic field, Bourland et al. (1990) determined the changing magnetic field strength (dB/dt) required to produce a sensory response. Twenty healthy adult volunteers were placed in the coil with the head outside. A capacitor discharge pulse delivered to the coil induced eddy-current pulses with durations of 1,189, 485, 382 and 228 μsec, the objective being the acquisition of a strength-duration curve for sensation. The resolution in determining threshold was 10%. The maximum pulse intensity available was 138T/sec. The intensity of these pulses was significantly higher than those used in present-day MRI. A strength (dB/dt)-duration (d) curve was obtained revealing that the shorter the duration of the pulse, the stronger the stimulus required. For the longest duration pulse (1,189 μsec), three subjects exhibited thresholds above 138T/sec, i.e. they could not be stimulated. Likewise seven subjects could not be stimulated with the 485 μsec pulse, 13 subjects could not feel the 352-μsec pulse and only one subject could feel the 228-μsec pulse with the maximum intensity available. No subject in the study reported the sensation to be uncomfortable. As stated earlier, the dB/dt values used in this study were significantly above those used in present-day MRI.

Budinger et al. (1990) measured the threshold for sensation in ten healthy volunteers using X and Z gradient coils outside the static magnetic filed. Sinusoidal current was used and for a frequency of 1,270 Hz, the threshold was 60T/sec.

Cohen et al. (1990) reported sensation and muscle contractions in two human subjects undergoing MRI. They stated: "When two human volunteers were imaged with magnetic field gradient dB/dt of 61 Tesla/sec RMS, the subjects reported, to our surprise, feeling muscular twitches synchronous with gradient pulses over repeated experiments. No adverse or sustained effects were seen. Experiments in a canine, intended to assess the safety of MR imaging with dB/dt of up to 66 Tesla/sec RMS, failed to induce detectable changes in the electrocardiogram or to show any signs of gross response to gradient pulsing". In analyzing these results, Cohen et al. stated: "Though these human data are anecdotal, they do offer some suggestion as to the threshold for direct magnetic stimulation. In prior studies made at 2.0 T with primary gradient field variations of up to 28 T/s RMS and cross fields of up to 18 T/s RMS, (note that the gradient geometry of that system was such that the cross fields were of smaller magnitude than the primary fields), no direct stimulation effects were reported (references and unpublished observations), following the collection of over 11,000 images on 60 volunteers and patients. Even when that system was operated at 51 T/s rms (primary fields) and 32 T/s rms (cross field) on two volunteers, no stimulation effects were observed. However, when that instrument was tested with primary gradient fields up to 57 T/s rms and cross fields to 35 T/s rms on two volunteers, one volunteer (who had been noting nasal congestion) described sensations in the nasal sinus area which were not, at that time, attributed to direct magnetic stimulation."

The report concluded with: "We believe the stimulation effects reported here to be below the level of medical significance though they offer tantalizing clues as to the lower boundaries of direct magnetic stimulation."

In a dog study the threshold for peripheral motor-nerve and phrenic-nerve stimulation were determined by Bourland et al. (1990) using a 24-turn coil surrounding the thorax. Single pulses ranging from 65 to 530 μsec were used and the resulting muscle contraction due to motor-nerve stimulation was recorded with a myograph. Phrenic-nerve stimulation was recorded by measuring the volume of air inspired for each Z-gradient magnetic stimulus. Strength (dB/dt)-duration (d) curves were obtained for motor and phrenic-nerve stimulation. The rheobasic dB/dt values for these nerves were about 300 and 1,100 T/sec respectively. Cardiac stimulation threshold was determined for the longest duration pulse. The report concluded: "Peripheral nerves exhibit the lowest threshold for stimulation. Greater fields stimulate the nerves innervating muscles of respiration. At still greater levels, cardiac arrhythmias can be induced. At 530 μs, the cardiac stimulation threshold is 2.6 times that of the nerves of respiration and 9.4 times that of peripheral nerves."

Although cardiac arrhythmias have never been reported with MRI, it has been desirable to determine the magnetic field strength that will produce an ectopic beat. With a view to creating a new closed-chest cardiac pacing method, Bourland et al. (1990) placed two, 30-turn, co-planar (pancake) coils on the canine left chest over the apex-beat area. It required a peak current pulse of 9,250 amps to evoke a cardiac contraction. The mean energy in the pulse was 11,850 joules, indicating that it is very difficult to evoke a cardiac contraction by a pulsed magnetic field. Because of the high energy required, a closed-chest, magnetic cardiac pacemaker is impractical with present knowledge of coil design and the use of a 120-volt power source.

It soon became obvious that it was desirable to know how much below cardiac stimulation threshold the presently used MRI units were operating. Thus several directed studies were undertaken to provide this information. Nyenhuis et al. (1991) compared the cardiac stimulation thresholds in dogs using X, Y and Z gradient coils. The amplitudes of the capacitor-discharge pulses delivered to them were adjusted to reflect the stimulating properties of a rectangular pulse. A cardiac threshold of about 3,000 T/sec was measured for a 530-μsec rectangular pulse. Cardiac stimulation with the transverse coils was not observed with the maximum available pulse intensity of 1,302 T/sec for a duration of 572 μsec. They estimated a cardiac threshold of 1,500 T/sec for the transverse gradient coils.

Conclusion

From the foregoing it is clear that extremely high values of dB/dt are needed to stimulate the heart. Experience has shown that there is a hierarchy of thresholds, the lowest being for sensory stimulation. The threshold for peripheral nerve stimulation, evidenced by muscle contractions lies above the sensory threshold. Well above nerve stimulation threshold is the cardiac stimulation threshold.

REFERENCES

Adrian, E.D. The response of human sensory nerves to currents of short duration. Physiol. 1919, 53:70–85.

d'Arsonval A. Dispositifs pour la mesure des courants alternatifs de toutes frequences. C R Soc Biol 1896; 2:450-51.

Ayers, G.M., S.W. Aronson and L.A. Geddes. Comparison of the ability of the Lapicque and exponential strength-duration curves to fit experimentally obtained perception threshold data. Australas. Phys. Eng. Sci. Med. 1986, 9(3):111-116.

Bernstein, T. Personal communication 1976.

Blair, H.A. On the intensity-time relations for stimulation by electric currents. J. Gen. Physiol. 1932, 15:177-185, 709-729,731-755.

Bourland, J.D., Mouchawar, G.A., Nyenhuis, J.A., Geddes, L.A., Foster, K.S., Jones,J.T. and Graber, G.P. Transchest magnetic (eddy-current) stimulation of the dog heart. Med. Biol. Eng. Comput. 1990, March:196-198.

Bourland, J.D., Nyenhuis, J.A., Mouchawar, G.A. et al. Physiologic indicators of high MRI gradient induced fields. Proc. SMRM, p. 1276, (WIP). San Francisco, CA 1990.

Bourland, J.D., Nyenhuis, J.A., Mouchawar, G.A. et al. Human peripheral nerve stimulation from Z-gradients. Proc. SMRM, p. 1157, (WIP). 1990, New York.

Budinger, T.F., Fischer, H., Hentschel, D. et al. Neural stimulation dB/dt thresholds for frequency and number of oscillations using sinusoidal magnetic gradient fields. Proc. SMRM., 9th Ann. Sci. Mtg., p 276. N.Y. Aug 18-24, 1990.

Chokroverty, S. Magnetic Stimulation in Clinical Neurophysiology. Boston 1989, Butterworths, 308 pp.

Cohen, M.S., Weisskoff, R.M., Rzedian, R.R. Sensory stimulation by time-varying magnetic fields. Magnetic Resonance in Med. 1990, 14:409-414.

Dalziel, C.F. Effects of electric shock on man. IEEE Trans. BioMed. Eng. 1956, 5:44-62.

Dominguez, G. and A. Fozzard. Influence of extracellular K concentration on cable properties and excitability of sheep Purkinje cardiac fibers. Circ. Res. 1970, 26:565–574.

Edelberg, R. Personal communication 1963.

Fozzard, H.A. and M.Schoenberg. Strength-duration curves in cardiac Purkinje fibers. Physiol. 1972, 236:593–618.

Gangarosa, L.P., Park, N.-H., Wiggins, C.A. and Hill, J.M. Increased penètration of non-electrolytes into mouse skin during iontophoretic water transport (iontohydrokenesis). Pharm. Exp. Therap. 1980, 212:377-381.

Geddes, L.A. Electrodes and the Measurement of Bioelectric Events. New York 1972, Wiley Interscience. 364 pp.

Geddes, L.A. and J.D. Bourland. Tissue stimulation: Theoretical consideration and practical application. Med. Biol. Eng. Comput. 1985, 23(2):131–137.

Geddes, L.A. The history of stimulation with eddy currents due to time-varying magnetic fields. In Magnetic Stimulation in Clinical Neurophysiology. Chokroverty, S. ed. Boston 1989, Butterworths, 308 pp.

Geddes, L.A., and Baker, L.E. Response to the passage of electric current through the body. Assoc. Adv. Med. Instrum. 1971, 5:13-18.

Koslow, M., A. Bak, and C.L. Li. C-fiber excitability in the cat. Neurol. 1973, 41:745–753.

Lacourse, J.R., Miller, W.T., Vogt, M. and Selikowitz, S.M. Effect of high-frequency current on nerve. IEEE Trans. Biomed. Eng. BME 32(1):82-86.

Lapicque, L. Definition experimental de l'excitation. C.R. Hebd. Seances Acad. Sci. 1909, 67(2):280-283.

Lapicque, L. L'Excitabilite en Function du Temps. 1926, Presses Univ. de France, Paris, 371 pp.

Lee, W.R., and Scott, J.R. Thresholds of fibrillating leakage currents along ventricular catheters. Cardiovasc. Res. 1973, 7:495-500.

Li, C-L., and A. Bak. Excitability characteristics of the A and C fibers in the perineal nerve. Exp. Neurol. 1976, 50:67–79.

Licht, S. Therapeutic Electricity and Ultraviolet Radiation. 2nd ed. Baltimore MD, 1967, 434 pp.

Maloney, J.M., Bezzant, J.L., Stephen, R.L. and Petelenz, T.J. Iontophoretic administration of lidocaine anesthesia in office practice. Dermatol. Surg. Oncol. 1992, 18:937-940.

Monsees, L.R., and McQuarrie, D.G. Is an intravascular catheter a conductor? Med. Electron. Data 1971, 12:26-27.

Notermans, S.L.H. Measurement of the pain threshold determined by electrical stimulation and its clinical application. Neurology 1966, 16:1071–1086.

Nyenhuis, J.A., Bourland, J.D., Mouchawar, G.A. et al. Comparison of stimulation effects of longitudinal and transverse MRI gradient coils. Proc. 10th Ann. Mtg. SMRM, p 1275, San Francisco.

Pearce, J.A., J.D. Bourland, W. Neilsen, L.A. Geddes, and M. Voelz. Myocardial stimulation with ultrashort duration current pulses. PACE 1982, 5:52–58.

Prevost, J.L. and Batelli, F. Death by electric current (constant). Comptes Rendus Acad. Seance (Paris) 1889, 128:842-844.

Ritchie, A. The electrical diagnosis of peripheral nerve injury. Brain 1944,

67:314–330.

Roy, O.Z., Scott, J.R., and Park, G.C. 60 Hz ventricular fibrillation and pump failure threshold versus electrode area. IEEE Trans. Biomed. Eng. 1976, BME-23(1):45-48.

Roy, O.Z., Park, G.C., and Scott, J.R. Intracardiac fibrillation threshold as a function of the duration of 60Hz current and electrode area. IEEE Trans. Biomed. Eng. 1977, BME-24(2):430-435.

Schaefer, D.J. Safety aspects of magnetic resonance imaging. In Biomedical Magnetic Resonance Imaging, Wehrli, F., Shaw, D. and Kneeland, B. Eds. New York 1988, VCH Publishers.

Schenck, J.F., Dumoulin, C.L., Souza, S.P. et al. Health and physiological effects of human exposure to whole-body 4 Tesla magnetic fields during magnetic resonance scanning. Proc. Soc. Mag. Resonance Imag. 1990, Aug. 18-24. New York, page 277.

Staewen, W.A., Mower, M., and Tabatznik, B. The significance of leakage currents in hospital electrical devices. Mt. Sinai Hosp. (N.Y.) 1969, 15:3-10.

Starmer, C.F., and Whalen, R.E. Current density and electrically induced ventricular fibrillation. Med. Instrum. 1973, 7(2):158-161.

Tyle, P. Iontophoretic devices for drug delivery. Pharmaceut. Res. 1986, 3:318-326.

Tyle, P. Iontophoretic Devices, Chap 14 in Drug Delivery Devices. New York, 1988 Marcel Dekker, 607 pp.

Underwriters Laboratories Standard for Electric Fence Controllers. ANSI/UL 69, July 10, 1979. U.L. 233 Pfingsten Rd., Northbrook, IL 60002.

Wadas, R.S. Biomagnetism. New York 1991, Ellis Horwood 170 pp.

Weinberg, D.I., and Artley, J.C. Electric shock hazards in cardiac catheterization. Circ. Res. 1962, 1:1004–1009.

Weiss, G.G. Sur la possibilite de rendre comparables entre eux les appareils a l'excitation. Arch. Ital. Biol. 1901, 35:413-446.

Whalen, R.E., Starmer, C.F., and McIntosh, H.D. Electrical hazards associated with cardiac pacemaking. Ann. N.Y. Acad. Sci. 1964, 111:922.

Chapter 3
RESPONSE TO LOW-FREQUENCY
ALTERNATING CURRENT
PASSING THROUGH THE BODY

INTRODUCTION

Response to the passage of low-frequency sinusoidal alternating current (AC) through the body depends on the current pathway, the current magnitude (amps) and the duration of current flow. However, it should be recognized that the sudden, unexpected encounter of low-intensity current can produce an injury by startling a subject and causing a loss of equilibrium, the injury being due to the fall, not the current intensity. In a typical accident situation, the voltage is usually known but the current and its duration of flow are subjects for speculation. However, it is possible to estimate the current from some data available on contact and body resistance; these factors are discussed in Chapter 6. The current pathway can often be inferred from the presence of contact sites. Jex-Blake (1913) introduced the concept of an entering and leaving site, being the analog of gunshot injury, in which the injury at the entering site is small; the exit wound is much larger. These terms are not appropriate for electrical injury; "contact sites" is the term used.

ENCOUNTERABLE SOURCES

The most commonly encountered alternating current derives from domestic outlets,which is 120-volt (rms), 60 cycles/sec (Hz) in the US and 120 or 240 volts (rms), 50 Hz in many parts of Europe. Voltages from 440 up to 7,500 are present in many factories. Short-distance transmission lines operate at 10,000–20,000 volts, while long-distance lines operate at 200,000 to 750,000 volts. Voltages of about 80 are used in AC arc welders.

Alternating current is described in terms of the root-mean square (rms) value for voltage and current. The rms value for voltage and current are $1/\sqrt{2}$ or 0.707 times the peak voltage or current. Figure 3.1 illustrates an analog record of two cycles of 60 Hz, 120 volts (rms). Note that the peak positive and negative voltages are $120\sqrt{2} = 170$ volts. The peak-to-peak voltage is $2 \times 170 = 340$ volts. The reason for using the rms notation derives from power considerations. In other words, the product of rms voltage and rms current is the power in watts which is equal to the same power as for a direct voltage and current equal to the rms values.

In aircraft there are engine-driven 400 Hz, 115/208 volt supplies.In automobiles with engine-driven alternators, the frequency range is about 40–200 Hz, depending on the engine speed. However, this voltage doesn't appear outside of the alternator housing because the rectifiers that are used to produce the direct current for battery charging are within the alternator housing; only direct current emerges. The high-frequency, short-duration pulses associated with ignition are discussed in Chapter 2.

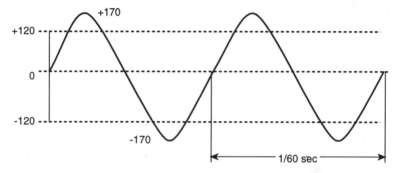

Figure 3.1   The relationship between rms voltage and peak voltage.

PATHOPHYSIOLOGICAL EFFECTS

As stated previously, the effects of low-frequency alternating current depend on the current pathway, magnitude and duration of current flow. However an additional factor, such as the manner in which the current flow is terminated, may be important in a given situation. In all cases, it is useful to note that the main hazards to the passage of alternating current relate to respiratory arrest, ventricular fibrillation, muscle contractions and burns.

As shown in Chapter 2, the stimulating capability of alternating current decreases with increasing frequency (strength-frequency curve). In other words, the higher the frequency, the more the current needed to stimulate excitable tissue. Before discussing the hazards of power-line (50–60 Hz) current, additional information will be presented on the effect of frequency.

IMPORTANCE OF FREQUENCY

In Chapter 2 it was shown that individual excitable tissues require more current for stimulation as the frequency is increased. We will examine a few situations in which sinusoidal alternating current of increasing frequency was applied to body-surface electrodes.

The vagus nerves run in the neck and innervate the heart, their function being to slow or stop the heart. Figure 3.2 shows the threshold sinusoidal current (rms) applied to circum-neck (N), circum-abdomen (A) and transchest (TC) electrodes required to produce slowing of the canine heart as the frequency was increased. Note 1) that less current was needed to slow the heart with neck-abdomen electrodes and 2) with both electrode configurations, more current was required for cardiac slowing as the frequency was increased above about 50 Hz.

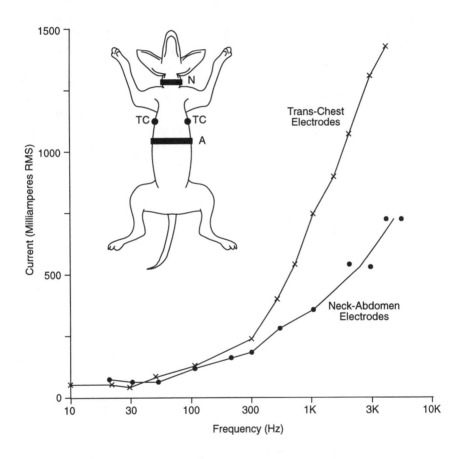

Figure 3.2. Average threshold current for vagal stimulation using trans-chest and neck-abdomen electrodes applied to dogs weighing 11–18 kg. Vagal stimulation was observed by slowing of the heart rate. (Redrawn from Geddes, L.A. and Baker, L.E. Med. Biol. Eng. Comput. 1969, 7:289–296.)

Lee (1961) drew attention to an important aspect of power-line current passing through the thorax via the hands which resulted in chest symptoms, such as tightness around the chest, difficulty in breathing and a sensation of strangulation. In these cases, the current held the subject to the current source by strong muscle contractions. Some subjects extricated themselves; others were removed by co-workers. None of these victims died, indicating that the level of current for this response is less than that required for ventricular fibrillation.

Figure 3.3 presents the threshold current for precipitating ventricular fibrillation with gradually increasing 60-Hz current applied to various electrodes on the bodies of dogs ranging in weight from 10 to 18 kg; lead 1 is right-forelimb to left forelimb; lead 11 is right forelimb to left hind limb; lead 111 is left forelimb to left hindlimb; lead aVR is left forelimb and hindlimb joined paired with the right forelimb; aVL is the right forelimb and left hindlimb joined and paired with the left forelimb; RA-RL is right forelimb to right hindlimb and LA-RL is left forelimb to right hind limb. As was found for cardiac slowing, the current required to precipitate ventricular fibrillation increased markedly as the frequency was increased above 100 Hz. Similar information was provided by Kouwenhoven et al. (1936), who, using interrupted direct current, found that the threshold current for fibrillation at 1,500 Hz was about 10 times that at 60 Hz.

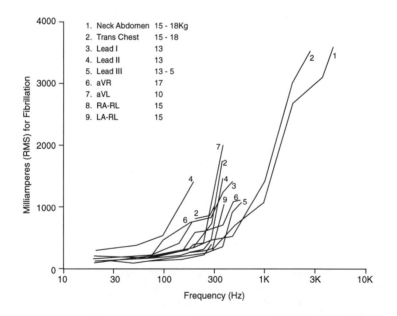

Figure 3.3. Threshold sinusoidal current for precipitation of ventricular fibrillation in the dog using various electrode locations. Ventricular fibrillation was precipitated by gradually increasing the current while monitoring the electrocardiogram and blood pressure. (Redrawn from Geddes et al. The Nervous System and Elect. Current. 1971, 2:121–129.)

CURRENT EXPOSURE TIME

As stated previously, the duration of exposure to power-line current is an important determinant of the response. Figure 3.4 presents the threshold 60-Hz current required for precipitating ventricular fibrillation in a 94-kg pony and 12 and 7.5-kg dogs. Note 1) that the current required for ventricular fibrillation with left-forelimb-to-left hindlimb electrodes increases with decreasing current exposure time, the long-duration asymptote being in the 1–5 sec range, and 2) more current is required for heavier subjects.

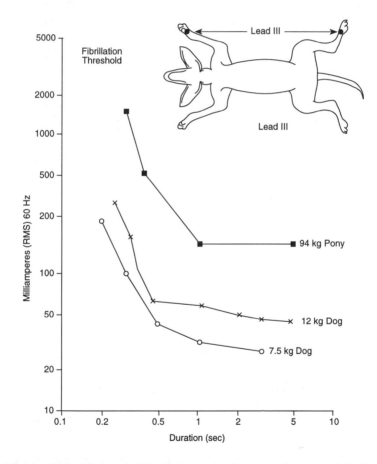

Figure 3.4. Current (60 Hz) required to produce ventricular fibrillation versus exposure time for lead 111 in animals of different weights. (Redrawn from Geddes et al. IEEE Trans. on Biomed. Eng. 1973, BME 20:465–468.)

EFFECT OF BODY WEIGHT AND CURRENT PATHWAY

Figure 3.5 presents the threshold current, for 5 seconds exposure time, versus body weight for precipitating ventricular fibrillation in subjects ranging form 0.5 to 200 kg; lead 1 is left forelimb to right forelimb; lead 11 is right forelimb to left hindlimb and lead 111 is left forelimb to left hindlimb. Note 1) that more current is needed to precipitate ventricular fibrillation with heavier subjects (in fact the needed current increases almost as the square root of body weight), and 2) less current is required in the head-foot direction (leads 11 and 111), compared to the trans-shoulder direction (lead 1). Essentially the same data were obtained by Ferris et al. (1936).

The data in Figure 3.5 can be used to estimate the head-foot (≥ 5 sec) current for precipitating ventricular fibrillation in a typical 70-kg adult. Using regression equations for lead 11 and 111, the 60-Hz, rms currents are 259 and 215 mA (rms) respectively, with an average of 237 mA.

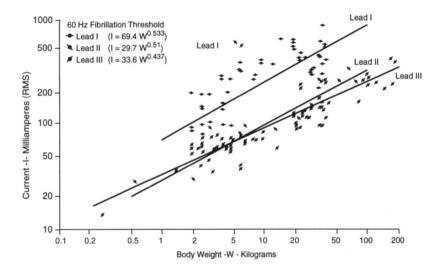

Figure 3.5.   Threshold 60 Hz current required for producing ventricular fibrillation in animals of different weights using 5 sec of exposure and leads 1, 11 and 111. (Redrawn from Geddes et al. IEEE Trans. on Biomed. Eng. 1973, BME 20:465–468. By permission.)

LOW AND HIGH-INTENSITY THORACIC POWER-LINE CURRENT

Although it is well established that there is a relationship between the threshold current and body weight for the induction of ventricular fibrillation with power-line current, the same type of current at a higher intensity can also achieve defibrillation. Beck et al. (1947) was the first to defibrillate the human heart electrically by applying 110 volts (rms), 60-Hz current for about 0.5 sec to transventricular electrodes. The current flow is estimated to be 3 to 5 amps rms (Geddes and Hamlin 1983).

With anterior chest electrodes, Zoll et al. (1956) were the first to defibrillate the human heart by applying 500–700 volts (rms) for about one half second to anterior-chest electrodes. The current that flowed was typically 10–15 amps (rms) for a 70-kg subject.

It may seem paradoxical that 60-Hz current can induce ventricular fibrillation as well as arrest it. The explanation lies in the intensity of the current. Recall that it takes only milliamperes to induce ventricular fibrillation with direct-heart electrodes. With such electrodes, defibrillation required several amps. Likewise, with current applied to thoracic electrodes, it required 250–600 mA (rms) to induce ventricular fibrillation; whereas defibrillation required 10–15 amps. In other words low current can induce ventricular fibrillation; high power-line current can defibrillate. This fact was known to Prevost and Batelli (1899) who were investigating the relative safety of direct and alternating current. In fact, their little-read report described the first successful ventricular defibrillation in animals. Using 45-Hz current, they wrote: "One can, in submitting the animal in which the heart is in fibrillary tremulation [fibrillation] by a current of low tension [voltage], see the heart re-establish ventricular contractions if one submits the animal to a current of high tension", [Author's translation].

The same message was given by Hooker et al. (1933) who used the dog and stated: "Alternating currents of five or more amperes when passed through the body for one-half to five seconds will not produce ventricular fibrillation. However, the usual house circuit (110 volts) similarly applied will invariably produce fibrillation because the current that flows presumably is not sufficient to inhibit the heart."

"One milliampere of current, applied directly to the ventricular musculature, is sufficient to cause fibrillation and the extreme ventricular apex is as sensitive as any other point on the ventricles.

"With the electrodes applied directly to the heart, currents of 0.4 ampere for five seconds will cause fibrillation and currents of 0.8 ampere or more will stop fibrillation. A current of 0.8 ampere will not induce fibrillation and a current of 0.45 ampere will not stop fibrillation. In the intact animal with the electrodes on either side of the thorax, the current spreads out over the body tissues. In order to obtain a sufficient current through the heart to arrest fibrillation, the countershock current must be increased to a value of at least four or five amperes."

The foregoing is consistent with the report of Emerson (1961) which stated that in Great Britain approximately two thirds of the electrical fatalities are associated with voltages below 250. This also may explain the reason for fatalities among power-line workers, some of whom succumbed, while others survived. When the use of artificial respirators was introduced for all electrical accidents, it was found that only those shocked with high voltage survived; those who succumbed, even with artificial respiration, had received low-voltage (110–220 volt) shocks. In probably most of the high-voltage cases, respiratory arrest occurred, and the application of artificial respiration rescued those whose ventricles were not in fibrillation.

ALTERNATING CURRENT APPLIED TO THE HEAD

Power-line alternating current has been applied intentionally to a pair of electrodes on the head for a fraction of a second to produce a convulsion, as in electroshock therapy. Higher frequency alternating current has been applied continuously to transhead electrodes on man to produce an anesthesia-like state (electronarcosis). The current intensity and duration of application, along with types of responses in both situations will now be described.

ELECTROSHOCK THERAPY

Electroshock therapy (EST) employs the passage of electric current through the head for several types of mental disorders described below. Various names have been used to describe the treatment; some of these are electroshock therapy, electroconvulsive therapy, electroshock treatment, brief-stimulus therapy and electroplexy (plexy-stroke).

The mental disorders treated with EST are schizophrenia, manic depression and severe depression with suicidal tendencies. Schizophrenia is an incapacitating disorder characterized by a general derangement of thought processes, manifested by a reduced interest in, and relationship with, the environment and its inhabitants. Emotion is lacking and ideation is simple. Behavior is reduced and usually stereotyped, often silly or bizarre. In some patients, there is a feeling of persecution and visual, auditory or gustatory hallucinations are not uncommon.

Manic-depressive (bipolar personality) behavior is characterized by excitement (manic phase) and also profound depression (depression phase). The manic is characterized by hyperactivity, often frenzied and incapable of sustained thought and easily distracted. Frequently such patients are destructive without malice. In the depression phase, the patient is inactive; usually the slightest activity is a great physical effort. Inattention, distrust, discouragement and fearful of impending disaster are common. A feeling of hopelessness and helplessness is common in depression. A variety of complaints are sometimes offered as being the cause of the state. The patient sleeps poorly, awakens early and may have suicidal tendencies because of a feeling of inadequateness.

The number of electrical treatments varies widely and depends on the type and severity of the disorder. Three to ten treatments may be given for depression, and from 12 to 25 for schizophrenia. The treatments may be administered as often as three times per week, or more often if the mental illness is severe. In annihilation therapy for refractory schizophrenia, several treatments may be given each day for several weeks.

The therapeutic benefit was believed to depend on the occurrence of a convulsion or grand-mal seizure, in which all skeletal muscles of the body participate and consciousness is lost immediately. The muscular contractions are first tonic (sustained), then clonic (jerking). Because of the intense muscular activity, bones have been fractured and joints dislocated. Laryngospasm, broken teeth, cardiac arrhythmias and apnea were reported in the early days of EST. To eliminate these undesirable side effects, medications such as a short-lasting anesthetic and a mild muscle relaxant are given; atropine is used to prevent the laryngospasm and reduce the cardiac arrhythmias. If a convulsion occurs due to passage of current through the head, on recovery, the patient is confused, disoriented and experiences a loss of memory for recent events (retrograde amnesia).

EST CURRENT AND ELECTRODE LOCATION

Cerletti (1950) introduced the use of sinusoidal alternating current for electroshock therapy. Depending on geographic location, 50 or 60 Hz sinusoidal current is used. Such current is applied to bitemporal electrodes for a few tenths of a second. The current intensity is in the vicinity of 200 to 500 mA (rms) when 120 volts are used. A 1:1 transformer provides an output which is isolated from ground.

Electroshock therapy was introduced using transtemporal electrodes, Friedman (1942) recommended placing one electrode on the vertex and the other over the temple, just above a line joining the ear and the eye. In the early unilateral studies, little regard was paid to the matter of hemispheric dominance. Later it was recommended that the temple electrode should be placed over the non-dominant hemisphere. In a right-handed person, this would be the right side of the head.

CURRENT DENSITY DISTRIBUTION

Ever since the early days of EST, there was in interest in the current pathway through the head. Mapping current density distribution in such a circumstance is not a trivial task. The current density at a point can be calculated from the voltage gradient (de/dx) and the resistivity (ρ) at that point. Note that the head is made up of tissues of differing resistivities and distributions, making it difficult to determine current density in different directions. Nonetheless, there have been a few studies which attempted to measure current-density distribution, particularly with regard to comparing data from the bilateral and unilateral electrode locations.

Smitt and Wegener (1944) used cadavers with bitemporal electrodes and biparietal electrodes. They found that the voltage gradient (voltage between two points divided by their separation) in the frontal-lobe region was higher with the bitemporal placement. With the biparietal placement, the current was more uniform in the brain. Of importance, they found that 90–95% of the voltage drop was across the scalp and skull, indicating that only a small percentage of the total current injected flowed through the brain. Lorimer et al. (1949) also carried out intracranial voltage measurements and concluded that current flow in the brain is not homogeneous, nor did it flow mainly in the cerebrospinal fluid (which has a low resistivity). Instead, they stated that it flowed along axonal bundles. With both bitemporal and vertex-lateral electrode placement, there was a high current density in the corpus callosum.

The live spider monkey was used by Hayes (1950) to determine the effect of scalp and skull on cerebral current flow. By applying electrodes to the scalp, then the skull and finally the brain, and measuring intracerebral voltages, he concluded that 80% of the applied current flowed along the scalp-skull pathway. Beyond the immediate vicinity of the electrodes, the current was relatively uniform. Finally, Weaver et al. (1976) carried out computer-based model studies directed toward determining the difference in current-density distribution with bilateral and unilateral electrodes. They equated the head to three concentric spheres 8, 8.5, and 9.2 cm in radius with resistivity (ρ) values of 220 ohm-cm for the brain, 17,700 ohm-cm for the skull and 222 ohm-cm for the scalp. The results of this model study showed that 1) the current density was highest in the frontal lobes, 2) with the unilateral placement, the current density was highest under the electrodes and near the surface of the simulated brain, along the imaginary line joining the electrodes. Comparing unilateral and bilateral electrodes, the current density with the latter is higher in all areas of the brain.

The fact that most of the current injected by a scalp-surface electrode does not reach the brain is easily explained by examination of the conducting properties of the tissues between the electrode and the brain cortex. Figure 3.6A illustrates the anatomic location and the resistivities of these tissues. Immediately below the electrode is the scalp with a relatively high resistivity (300–1,000 ohm-cm), below which is the high-resistivity (5,000–15,000 ohm-cm) skull, under which is the cortex with a typical resistivity of 500 ohm-cm. Figure 3.6B illustrates the current flow, showing that the high resistivity skull causes much of the current to flow extracranially.

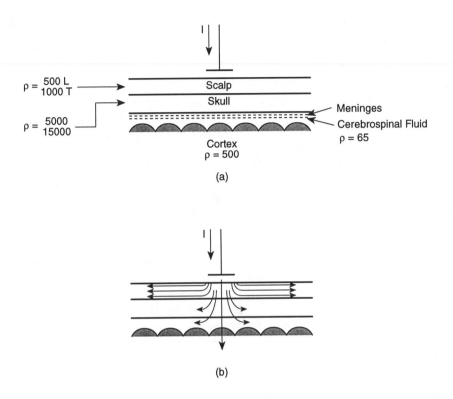

Figure 3.6. Electrode on the scalp showing the resistivity (ρ) values of the tissues thereunder (A) and current flow (B); L = lateral, T = transverse. (Redrawn from Geddes, L.A. Neurosurg. 1987, 20:94–99.)

HISTORICAL BACKGROUND OF EST

The use of alternating current applied to the head for therapeutic purposes is due to Cerletti, an Italian psychiatrist. In 1950 he described his early research carried out in the late 1930's. He had been using drug-induced convulsions in patients while in Genoa. He began to ponder the use of electric current as a convulsant because it had been shown by physiologists in the late 1800's that low-intensity, pulsatile electric current applied to the brain could produce epileptic seizures in animals. Accordingly, Cerletti started a study using dogs as experimental animals and applied power-line (50 Hz) alternating current to electrodes in the mouth and rectum. Convulsions were produced with a few tenths of a second of current flow. Subsequent examination of the brains revealed no histologic damage. In 1938 he learned that electric current was being used in the Rome slaughterhouse to stun pigs. Cerletti went there to see the procedure for himself. To his surprise, he found that the power-line current was applied with a hand-held pincer electrode assembly which delivered the current to the sides of the head. (This method is still in use to stun animals throughout the world.) When the voltage was applied, the pig convulsed and, while it was unconscious, the butcher made a deep slash in the neck causing profuse bleeding. Cerletti at once recognized the same type of convulsion seen in his dogs and surmised that the pigs died of exsanguination, rather than due to the electric current. Cerletti then obtained permission from the director of the slaughterhouse to conduct convulsive experiments on the pigs using power-line current applied to electrodes at various sites. The effect of current duration was also investigated. These studies clearly demonstrated the reliability of electric current to induce a convulsion, as well as the fact that after the convulsion, in all cases, the pigs recovered, even after being convulsed several times. In these studies it was found that 120 volts applied for a few tenths of a second was adequate to induce a convulsion.

Cerletti soon had the opportunity to apply electroconvulsive treatment to a 40-year old male schizophrenic who talked in a incomprehensible Italian gibberish. Transtemporal electrodes were applied and secured with an elastic band. Seventy volts were applied for 0.2 seconds. The patient stiffened and lost consciousness for a brief time. A later application of 110 volts for 0.5 second produced a typical epileptic convulsion. When the patient awakened, Cerletti inquired "What has been happening to you?" The patient answered in clear Italian, "I don't know; perhaps I have been asleep."

Following this remarkable demonstration, many power-line current electroconvulsive therapy devices became available commercially. Most had a voltage control and delivered current for a few tenths of a second. The availability of such equipment gave psychiatrists a new non-drug method of treating schizophrenia and severe depression. The introduction of other current waveforms soon followed. Nowadays, premedication is given to prevent the convulsion and strong muscular contractions.

The most important fact emerging from the foregoing narrative is that the application of 110-volt power-line current to the head does not result in death. Convulsions have resulted and there is the risk of respiratory arrest, which seems not to have been a prominent finding with transcranial power-line current.

ELECTRONARCOSIS

The prolonged passage of a wide variety of low-intensity pulsatile current, sometimes combined with direct current, through the head produces an anesthesia-like state which is promptly reversed as soon as the current flow is stopped. From the 1930's this method has been investigated (see review by Geddes 1965) and became very popular in the mid 1960's in animal research. The use of low current applied to electrodes over the eyes and back of the head has been used to produce sleep in man.

Attractive as electronarcosis appears, there are side effects and difficulties, the most serious of which is fading, i.e. the subject becomes very restless and appears to become less narcotized, despite continued current flow at the same intensity. Nonetheless electronarcosis can be used for short-duration surgical procedures. (Geddes et al. 1964). The following is a review of the studies in which alternating current has been used to produce an anesthesia-like state. Van Harreveld et al. (1943, 45, 47, 49) demonstrated that 50-Hz sine wave current applied to the heads of dogs produced narcosis. In another investigation they produced electronarcosis by suddenly applying 300 mA of 60 Hz to trans-temporal electrodes. Following the 30-sec induction phase, they reduced the current to 30–60 mA, the level they found best for narcosis. Like others, they noted two states, narcotic and kinetic. They also investigated other frequencies and found that the lowest current for the maintenance of electronarcosis was obtained at 100 Hz. Frostig et al. (1944) experimented with 60-Hz current suddenly applied to the heads of dogs. They used 300 mA for induction and then reduced the current by 50 mA to restore respiration. Higher frequencies were used by Denier (1938), who eliminated the strong muscular contractions in rabbits with currents of 85–120 kHz turned on for 1.5 msec and off for 13.5 msec. With electrodes placed between the eyes and over the sacrum, a current of 80–140 mA was employed. Knutson (1954) passed 700–1,500 Hz sine wave currents through the heads of dogs, to discover all of the responses referred to earlier. He noted that a high initial current produced convulsions which could be avoided with a gradual increase of the current. He also found that if the current was increased slowly, the animal became excited and, to maintain the animal under anesthesia, the current had to be raised to levels higher than that required with sudden application. Two years later, Knutson et al. (1956) reported additional studies on dogs using 700-Hz sine wave currents of 50–100 mA, applied to transtemporal electrodes. Using muscular relaxants and parasympathetic blocking drugs, they kept the animals narcotized for 3 hours. After gaining experience with the 700-Hz current applied to animals, they electronarcotized 5 human subjects using trans-temporal electrodes. The initial current employed varied between 135–150 mA. To maintain the anesthetic state, only a slight increase in current was needed. The subjects, who were maintained under electronarcosis for 12–32 min, exhibited hypertension and tachycardia.

Hardy et al. (1959) applied 700 Hz to the heads of premedicated dogs. In different groups of dogs they measured a variety of biochemical events comparing electronarcosis with local anesthesia, ether and a barbiturate. On the basis of their measurements they concluded that electronarcosis constituted a stress greater than accountable for by the surgical procedure. Having gained considerable experience with electronarcosis in animals, Hardy et al. (1961) used the same type of current for surgical procedures in patients.

A study employing various frequencies of sine wave current was carried out by Wulfsohn and McBride (1962). Starting with 700 Hz applied to electrodes placed across the head, they anesthetized 89 animals (rats, rabbits, guinea pigs, dogs and baboons). They found it necessary to premedicate all animals and observed that with rapid induction, in the frequency range 60–2,000 Hz, 1,500 Hz appeared to produce the best anesthesia. The currents employed in the dogs varied between 50–100 mA, while that used for rabbits was 20–35 mA.

DISCUSSION OF CRANIALLY APPLIED CURRENT

The physiological responses to a cranially injected current are many. Despite the fact that much of the current is extracranial, current reaches the cortex and can produce strong muscular contractions by stimulation of the motor cortex, as well as skin, auditory and visual sensations. Current reaching the medullary respiratory center at the base of the brain produces respiratory depression or arrest. Likewise, current reaching the medulla can produce vagal slowing of the heart, defecation and micturition. Despite these effects, it is clear that the many animal and human studies carried out show that if respiratory arrest is avoided or treated by artificial respiration, survival is assured.

In electroshock therapy the current flow is short, being less than 1 sec. Unless the patient is premedicated, there are strong muscle contractions, a convulsion and respiratory depression. With electronarcosis, the current is lower, but maintained, and the induction current is higher than the maintenance current. Many of the side effects of electroshock therapy may be present. In all cases where current has passed through the head, prompt artificial respiration is essential. If the subject is also pulseless, chest compression is needed to provide circulation until the heart starts to beat.

HIGH-VOLTAGE INJURY AND BURNS

High voltage is understood to be a voltage above 600. The injuries sustained over a range of power-line voltages are different. Because ventricular fibrillation is associated with voltages below about 300, this voltage could be used to provide a hazard boundary range. As with any electrical accident, the type of injury will depend on the voltage (which determines the current flow), the current path and the duration of exposure to the current.

In addition to skeletal muscle contraction, respiratory arrest (if the current reaches the respiratory center at the base of the brain) and potential cardiac arrhythmias, high voltage can produce a first, second or third-degree burn at the contact sites. A first-degree burn is skin reddening, as in mild sunburn. A second-degree burn is associated with blistering. If not too deep, second-degree burns heal by themselves. A third-degree burn involves damage to all of the skin layers and is often called a full-thickness burn. Often charring of the tissue is present. Third-degree burns require skin grafting.

Many animal studies have been carried out to describe the nature of high-voltage injuries. Usually the pig or hog is used because its skin is a good analog of human skin. Sances et al. (1981) measured the rise in subcutaneous tissue temperature 15 cm distant to the contact site on one hindquarter and toward a 40 × 60 cm plate under the other hindquarter. Using 2,100 volts, a 5-cm length of #2ACSR wire was brought to the skin surface while current and skin temperature were recorded. Just before skin contact, there was an arc to the skin. With continued contact the skin broke down with blistering and a full-thickness burn resulted, accompanied by tissue splattering and arcing. Figure 3.7A illustrates the subcutaneous temperature 15 cm from the contact site, revealing that in about 300 msec, the tissue temperature reached 100°C (boiling point). Figure 3.7B illustrates the current which varied considerably until the point of arcing (x), which occurred at 250 msec. In another study, Sances et al. (1981) placed a 5-cm metal disk on the hindlimb of a hog. A voltage of 40 was applied to the disk, which was then removed and a thermogram was made of the skin temperature. Figure 3.8 shows the outline of the electrode on the skin and the skin-surface temperature just after removal of the electrode. As expected, because the current density is highest under the electrode perimeter, the heating is maximum at this site.

In summary, high-voltage accidents may well involve many of the responses to low-voltage current. In addition, first, second or third-degree skin burns may be present, depending on the current intensity and duration of flow. Arcing, tissue blistering and charring are usually associated with high-voltage injury.

LEGAL ELECTROCUTION

Death by the passage of electric current through the body is designated "electrocution." The method was first adopted by the State of New York in 1888. The first criminal was electrocuted in 1890; details of this event have been reported by Bernstein (1973). In legal electrocution, the subject is seated in a wooden chair with both arms and legs strapped to the chair, as shown schematically in Figure 3.9A. One electrode is applied to the shaved head, the other to the left shin. The voltage-time program is different in different states, the method consisting of the sudden application of about 2,000 volts and consciousness is lost instantly with strong muscular contractions occurring at the instant of current flow. Both direct and alternating current have been used in legal

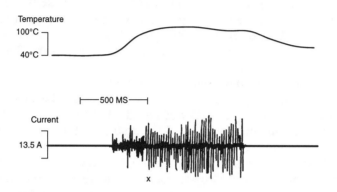

Figure 3.7.   Temperature and current data obtained with a #2ACSR wire (5 cm long) in contact with the hindquarter of a hog. The subcutaneous temperature probe was 15 cm proximal to the contact point. The dispersive electrode (40 x 60 cm) was on the opposite hindquarter. Arcing (x) started 250 msec after application of the current (From Sances, A.D. et al. IEEE Trans. Power Appar. Systems 1981, PAS100: 552–558, by permission).

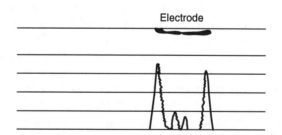

Figure 3.8.   Electrode (5 cm diam.) on skin (A) and thermogram of the skin surface (B). Each line represents 1°C. Note the high temperature under the electrode perimeter (Redrawn from Sances, A.D. et al. Journ. Trauma 1981, 21(8):589–597 by permission of the author and publisher, Williams & Wilkins, Baltimore, MD).

electrocution; however, the majority of applications use alternating current. The voltage is applied for a time from a few seconds to about 10 sec. In a typical case, about 40 A of current flows; then the voltage is reduced to about 500 and maintained at this value for about 30 seconds. From this point on the programs of raising and lowering the voltage vary.

In most cases, the total period of current flow is about 2 min. Figure 3.9B illustrates several voltage-time protocols.

To ensure that recovery is not possible, the law provides for an immediate postmortem examination. The pathologic findings from electrocuted criminals have been reported by MacDonald (1892), Spitzka and Radash (1912), Jex-Blake (1913), Jaffe (1928), Langworthy (1930), Pritchard (1934), Hassin (1933) and Alexander (1938). These reports indicate that there is virtually complete thermal destruction of the central nervous system in legal electrocution; brain temperatures as high as 140°F have been measured. Ventricular fibrillation has rarely been reported. There is general agreement that in legal execution, consciousness is lost instantly and death is due to thermal destruction of the central nervous system.

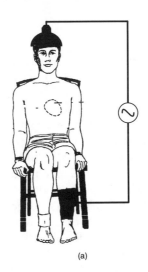

(a)

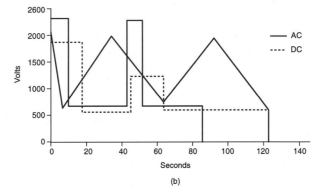

(b)

Figure 3.9. In A is sketched the manner of applying current in legal electrocution. B shows how the voltage is applied.

EUTHANASIA

Use of the electric chair has diminished in many states and the death sentence is carried out chemically. In a typical case, the condemned subject is placed on a gurney with the arms restrained. An intravenous line is established and connected to a programmable intravenous infusion device equipped for the injection of a barbiturate and potassium chloride. In some protocols, a muscle relaxant is included. First, the barbiturate is injected, producing anesthesia and respiratory depression or arrest. Cardiac action is then arrested by the injection of potassium chloride, which depolarizes cell membranes, rendering them inexcitable. The heart usually arrests in diastole (Hoff 1955); occasionally ventricular fibrillation occurs. In either case, all cardiac pumping ceases.

LEAKAGE CURRENT

The term leakage current is used to identify an undesired current flowing through a subject from a power-line operated device. To understand how leakage current arises, it is useful to recall that domestic power outlets have three terminals and that one side of the power line is grounded. Figure 3.10 illustrates the conventional wiring of the standard (single-phase) 3-prong receptacle which accepts a plug with two blades and a rod. The "hot" or ungrounded side of the power line is color-coded black, the return or "cold" grounded conductor is color-coded white. Current is delivered to a device (R) connected to these two conductors as shown in Figure 3.10. Power current is never conducted by the green or grounding conductor connected to the circular opening on the receptacle; this connection serves to ground the metal housing of a device plugged into the three-conductor receptacle.

Figure 3.10 shows a power-line operated device of resistance R in a metal housing which is not connected to the circular grounding conductor. Any two conductors separated by an insulator constitute a capacitor. Therefore the metal components of the device of resistance R and the metal housing constitute a distributed capacitor.

If a subject touches an ungrounded metal housing and ground, as shown in Figure 3.11A, current will flow from the hot side of the power line through the distributed capacitance (C_1) and through the subject to ground. The amount of leakage current flowing through the subject will depend on the voltage (E), the capacitance (C_1) and inversely with the resistance of the subject to ground. Obviously the subject will feel a mild or strong leakage-current shock, depending on the magnitude of C_1, all other quantities being the same. It should be noted that there may be leakage resistance due to faulty insulation. In this case, C_1 would be shunted by the leakage resistance and the leakage current would be larger.

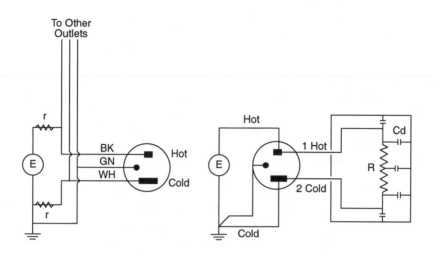

Figure 3.10. *Current is delivered to a device (R) via the "hot" (ungrounded) and "cold" (grounded) power conductors. When the power-consuming device (R) is within a metal housing, there is distributed capacitance (Cd) to it.*

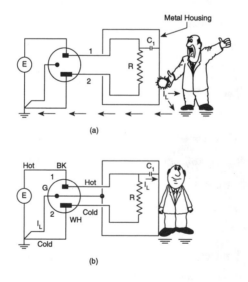

Figure 3.11. *A power-consuming device (R) within a metal housing has a capacitance (C_1) to the hot side of the power line to the housing (A). If a subject contacts the housing and ground, a leakage-current shock is perceived. With the metal housing connected to the ground (G) via the rod on the 3-prong plug; the housing will have no potential with respect to ground and the subject will receive no leakage-current shock (B).*

If the grounding (circular) terminal of the receptacle is connected to the metal housing by the three-prong plug, as shown in Figure 3.11B, the housing will be at ground potential and the subject in contact with the metal housing will receive no shock. To illustrate the magnitude of leakage current, the author connected a 5,000-ohm resistor in series with an AC microammeter and connected this circuit between the exposed, ungrounded metal housings of different appliances and ground. Table 3.1 presents a summary of the leakage currents in microamperes (rms). Recall that in all cases, the grounding conductor in the three-prong plug was disconnected to reveal the leakage current. When the housing was grounded as shown in Figure 3.11B, the leakage current was zero.

Table 3.1: Leakage currents to nearby waterpipe ground
(measured through 5,000 ohms)

Item	Maximum, μA
Electric Typewriter	4–20
Hot plate	13
Coffeemaker (large)	240
TV monitor	1000
Refrigerator	70
Drill Press	100
Small Centrifuge	1
Thermal Bath	20
Intercom	5
Oscilloscope	60
Physiograph	140
Coffee Pot	600
Capacitor Checker	1000
Tube Checker	50
Flowmeter - case	55
Blood flowmeter - probe	50

LEAKAGE-CURRENT TESTING

From the foregoing, it is clear that there is a need to establish test procedures to quantitate the 60-Hz leakage current that could flow through a subject connected to a power-line-operated device. Desirable as this is, there is no unanimity among the various standards-promulgating groups regarding the most representative electrical circuit that can represent a worst-case condition for a subject in contact with ground. Nonetheless, the differences between the standards groups is one of degree, rather than kind.

Although many leakage-testing instruments can be purchased which measure power-line leakage current, as well as leakage current due to high-frequency generators (such as radio and television stations and electrosurgical instruments), it is useful to have a simple tester for 60-Hz leakage current. Such a device is illustrated in Figure 3.12A and is very useful for testing a variety of power-line operated devices.

The leakage tester shown in Figure 3.12A consists of a three-prong power-line plug that is connected by three switches to a three-prong receptacle to which the device to be tested is connected. The tester permits reversing the power line, opening the ground connection, and opening either side of the power line connections. The leakage current is caused to flow to ground from the device being tested through an impedance (Z), which simulates the subject and the contact to ground. Associated with Z is the leakage current (I_L).

The simplest equivalent for Z (Figure 3.12B) is a 500 or 1,000-ohm resistor. The 500-ohm value was derived from the 60-Hz impedance of the body in good contact with ground (see Chap. 6). The 1,000-ohm value was recommended by the National Fire Protection Association (NFPA, 76BM-3042), which represented the impedance of a typical AC microammeter. With a simple resistor (R) used to simulate the subject, the leakage current is simply E, the voltage across the resistor, divided by the resistance (R) in ohms. A battery-operated AC voltmeter with a high input impedance must be used to make this measurement.

The American Association for the Advancement of Medical Instrumentation (AAMI) developed a standard circuit for testing leakage current. Figure 3.12C illustrates the circuit which exhibits a decreasing impedance with increasing frequency, as shown in Figure 3.12E. At d.c., the impedance is 1,000 ohms; the impedance is 10 ohms at infinite frequency. The impedance at 60 Hz is a fraction of a percent below 1,000 ohms.

The leakage current is determined by measuring the voltage (E) across the circuit using a voltmeter with a high input impedance or an oscilloscope, with due precaution been taken to avoid ground loops. The leakage current I_L = E/Z, where E is the measured voltage and Z is the impedance at the frequency of measurement. Note that for the same leakage current, the measurement voltage (E) will decrease as the frequency of the leakage current increases.

Figure 3.12D illustrates the circuit recommended by the Canadian Standards Association (C 22.2, 1979). The impedance-frequency characteristic of this circuit is illustrated in Figure 3.12E. The impedance decreases only slightly with frequency, starting out at 1,000 ohms at 0 Hz and decreasing to 993 ohms at infinite frequency. However, it is clear that the measured voltage (E) decreases with increasing frequency.

It is noteworthy that the American (AAMI) equivalent (Figure 3.12C) and the Canadian (C22.2) equivalent circuits (Figure 3.12D) are different in configuration, and have different impedance-frequency characteristics (Figure 3.12E). It is important to recognize that it is the measured leakage current that is important. In Figure 3.12F the voltage for 10 μA of leakage current is plotted versus frequency for both circuits. Although the two circuits are different, the frequency dependence for the measured leakage current is quite similar.

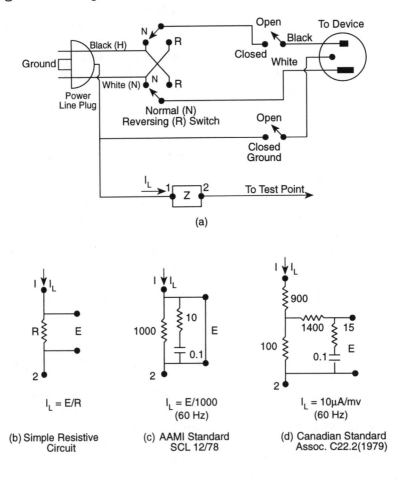

(a)

(b) Simple Resistive Circuit
$I_L = E/R$

(c) AAMI Standard SCL 12/78
$I_L = E/1000$
(60 Hz)

(d) Canadian Standard Assoc. C22.2(1979)
$I_L = 10\mu A/mv$
(60 Hz)

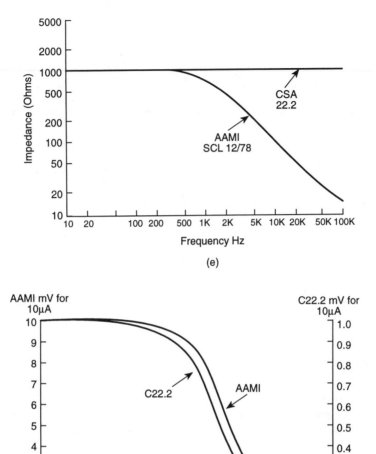

(e)

(f)

Figure 3.12. Leakage-testing circuit (A) in which Z represents the equivalent circuit. In B, Z is represented by a 500-ohm resistor. The AAMI standard (SCL - 12/78) is shown in C and the Canadian Standard Association circuit (C222.2, 1979) is shown in D. The impedance-frequency curves for C and D are shown in E. In F are shown the voltage-frequency curves for 10 A flowing through circuits C and D.

The leakage tester in Figure 3.12 permits performing eight leakage tests on a single test site in rapid succession by merely manipulating the switches in succession. Table 3.2 presents a document that can be used to log leakage currents found using this tester.

Table 3.2: Leakage Data

Test Conditions				Test-site Leakage Current				
Reversing Switch	Ground Switch	Line Switch	To Chassis	1	2	3	4	5
Normal	Closed	Closed						
Normal	Open	Closed						
Normal	Closed	Open						
Normal	Open	Open						
Reversed	Closed	Closed						
Reversed	Open	Closed						
Reversed	Closed	Open						
Reversed	Open	Open						

GROUND-FAULT CIRCUIT INTERRUPTER

A ground-fault circuit interrupter (GFCI) is a device that automatically arrests the flow of current to an appliance when the current in one of the power conductors exceeds that in the other. It should be obvious that when any appliance is plugged into the power line, the current (I) is the same in both conductors (1,2), as shown in Figure 3.13A. If there is a conducting path to ground and a fault current (I_f) flows in it, the current in the two power conductors (1,2) will differ by an amount equal to the ground-fault current. When a GFCI is interposed between the power outlet and the appliance, as shown in Figure 3.13B, a ground-fault current (I_f) can be detected and current flow arrested, as shown in Figure 13B. Suppose that a ground fault, represented by a resistance R_f in Figure 3.13B, is present. In such a case, the current in conductor 1 will be larger than the current in conductor 2 by an amount equal to the ground-fault current I_f. This inequality in current in conductors 1 and 2 is detected by a transformer (T) which has three windings, two for current and one for voltage (e). The connections are such that current flows in one direction in one current winding and in the opposite direction in the other current winding. Thus, the magnetic field in the core of the transformer (T) is zero when the currents are equal. In such a case there is no voltage (e) appearing across the voltage winding which is connected to an amplifier (A) and a latching relay with contacts in the normally closed position as shown; therefore current flows to the appliance. However, when the ground fault occurs there is a net alternating magnetic field in the core of the transformer and a voltage appears across the voltage winding (e). This voltage, after amplification (A), activates the latching relay which arrests the current flow to the appliance. The relay contacts stay in a locked-open position so that manual resetting is required to re-energize the appliance. If the ground fault has not been removed, the latching relay cannot be reset. Current can only be delivered to the appliance when $I_1 = I_2$, that is, when there is no excess flow of current in one of the two conductors to the appliance.

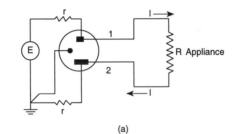

(a)

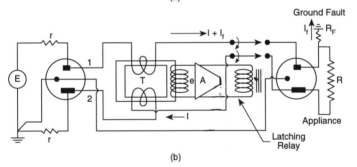

(b)

Figure 3.13. In A is shown a customer device (R) connected between the hot and cold sides of the power line and the current I is the same in conductors 1 and 2. In B is shown a GFCI and a fault to ground (R_f) resulting in a fault current (I_f) flowing through the fault. In this case the current in conductor 1 is $I + I_f$; the current in conductor 2 is I.

The speed with which a GFCI operates depends on the magnitude of the fault current. Figure 3.14 illustrates the response (trip) time for a typical unit. Observe that with a leakage current of 5 mA, which can be perceived easily, the GFCI arrested the current flow in 0.1 sec, i.e. when 6 cycles of 60 Hz current had flowed. The NFPA (76 BT-20641) recommends that the performance of a GFCI for hospital use should have an interrupt time of 25 msec for a 2 mA leakage current.

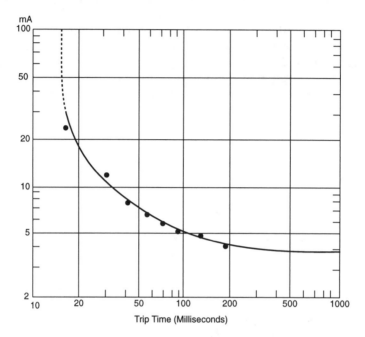

Figure 3.14. Ground-fault current (mA rms) versus trip time for a ground fault circuit interrupter (GFCI). From Health Devices, Special Edition 1972–3,2:24 pp.(Reprinted with permission of ECRI Plymouth Meeting, PA.)

ISOLATED POWER SUPPLY

When it is desired to reduce the electrical hazard due to leakage current, an isolated power supply is used. Such a power source is one in which neither side of the power line is connected to ground. An ideally isolated system would exhibit an infinite impedance from each side of the line to ground. Consequently, no current would flow in a conductor if it joined either side of the power line to ground. Obviously, owing to distributed capacitance, practical isolated power systems do not attain this ideal goal.

Isolation in a power system is achieved by the use of a transformer in which the secondary (output) winding has a low capacitance to ground, there being no conductive path to ground. Voltage is induced in the secondary by magnetic coupling which does not constitute a conductive path.

Figure 3.15A illustrates such an isolated power system. Note that between the primary (input) and secondary (output) winding there is an electrostatic shield which is grounded; therefore, there is a distributed capacitance C_1, C_2 between the secondary winding and ground, as shown in Figure 3.15B. In a well-designed and well-constructed isolation transformer, the capacitance is small and the insulation resistance is high; these two quantities define the quality of the isolation.

The ungrounded, that is, isolated, power distribution system is common in some hospital areas. The standards of performance for isolated power systems have been revised many times and the present recommendation appears in the National Electrical Code (Article 617).

The NFPA (76BM) recommends that a different color code be used to identify the conductors in an isolated power system; orange and brown are used for the isolated output and green is used for ground.

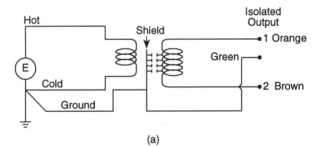

(a)

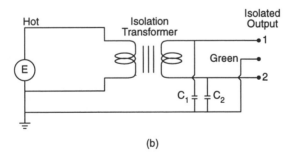

(b)

Figure 3.15. The use of a shielded transformer to provide an isolated power supply (A), and the capacitance (C_1C_2) to ground that characterizes the isolation (B).

EXTENSION CORD

Figure 3.16A shows a typical 3-wire extension cord used with the two-blade and-rod receptacle. Of major importance is the continuity of the grounding conductor connected to the rod at the male plug and to the conductor in the circular opening at the female end. A break in this grounding conductor can easily go undetected because the appliance connected by the extension cord will function normally. However, its metal housing will not be grounded. Figure 3.16B is a radiograph showing a broken connection between the grounding pin and the grounding conductor.

It is particularly unwise to use an extension cord on an isolated power supply because the distributed capacitance (C in Figure 3.16 to ground in the extension cord) diminishes the isolation provided by the isolating transformer. The capacitance of a typical 10-foot extension cord was measured by the author to be 250 $\mu\mu f$ between each conductor and the grounding (green) conductor. If an extension cord must be used, it should be placed ahead of the isolation transformer, the latter being placed as close as possible to the appliance to be isolated.

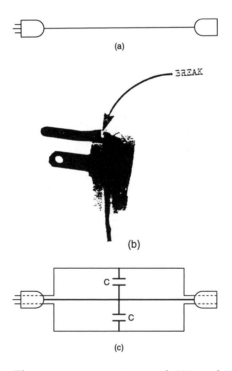

(a)

BREAK

(b)

(c)

Figure 3.16. Three-wire extension cord (A) and its electrical equivalent (C) identifying the capacitance (C) to ground from each conductor. B shows an X-ray photograph of a broken connection to the grounding pin.

LINE-ISOLATION MONITOR

A line-isolation monitor (LIM) is a device that measures the isolation provided by an isolated power supply, such as that shown in Figure 3.15. There are many sophisticated types which can detect symmetrical and asymmetrical faults to ground from the isolated conductors of an isolated power supply. Figure 3.17A illustrates the principle employed in a line-isolation monitor. The switch S connects a current-measuring device (I, Z) first to one of the isolated conductors (1); then to the other (2) in rapid succession. With no fault to ground to either conductor, the currents through Z will be equal and very small and relate to the magnitudes of C_1 and C_2 of the isolated lines.

With a ground fault, (R_f) as shown in Figure 3.17B, the current indicated when the switch is in position 2 will be larger than when the switch is in position 1. The inequality of currents indicates the presence of a fault. In a line-isolation monitor, the switch S is a solid-state device.

To detect R_f the LIM must draw a small current by completing a circuit to ground from the unfaulted side of the power line to ground. LIMs typically have an internal impedance (Z) of many hundred thousand ohms. It should therefore be obvious that the addition of an LIM to an isolated power system slightly reduces the degree of isolation. Sometimes small switching transients, due to operation of the switch (S), can be detected on the isolated line.

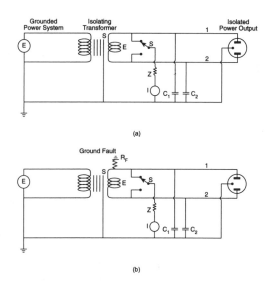

Figure 3.17. Principle used in a line-isolation monitor (A). With a fault to ground (R_f), the currents sampled when the switch (S) is in positions 1 and 2 will be unequal (B).

Many LIMs are equipped with both visible and audible alarms for excessive fault current, that is, when a fault impedance lower than a predetermined value develops. Modern LIMs will detect both symmetrical and asymmetrical faults to ground in an isolated power system. The NFPA (56A, 3344) recommends that "an alarm is activated, when line-to-line ground impedances exits that would allow more than 2.0 milliamperes in a direct connection between either isolated conductor and ground."

A simple and inexpensive LIM, shown in Figure 3.18, consists of two neon lamps with series resistors (r) connected across the isolated power supply; the connection between the two neon lamps is grounded, as shown. In normal operation with no ground fault, both neon lamps are lit. If either isolated conductor has a fault to ground, one of the neon lamps extinguishes. Practical as this monitor is, it cannot detect a symmetrical fault to ground.

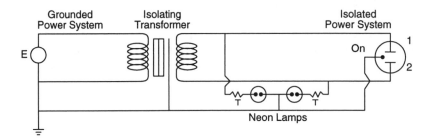

Figure 3.18. Neon-lamp, ground-fault indicator. With no fault, both neon lamps glow. Grounding either side of ht isolated power system will extinguish one neon lamp.

STRAY VOLTAGE

Stray voltage is a term that is not defined precisely; it pertains to an environmental voltage between two contact points on a human or animal that results in a mild shock. Typically a stray voltage is on the order of 10 volts or less. A source of stray voltage is current flow in the multigrounded neutral of a 60-Hz power line. Stray voltage has attracted attention, particularly on farms where animals are raised. Mild shocks received by cows result in reduced or variable milk production. According to Appleman, recognition of the behavioral response in animals to what we now call stray voltage dates from 1948 in Australia where stray current resulting from electrical equipment in the milking area may have affected cows negatively. A similar event was reported in New Zealand in 1962, and the first cases of stray voltage were reported in the US in 1967 and in Canada in 1975.

The concept of stray voltage is illustrated in Figure 3.19a in which power is delivered from a substation (AB), via conductors AC and BD to a load (R) at the customer site. The power source (E) at the substation is grounded by a long (8-foot) metal rod driven into the ground; the same practice is carried out at the customer site. Power is delivered by the ungrounded (hot) and grounded conductor, designated the PRIMARY NEUTRAL in this two-conductor system. Assume that there is a 5% voltage drop in each of the current-carrying conductors; therefore the voltage across the customer's load (R) is 0.9 E. Note that there will be a voltage amounting to 5% of E between the two ground rods at B and D, which will cause current to flow in the ground. This current flowing in the ground produces the stray voltage, as can be demonstrated by driving a metal rod X into the ground near D. If an alternating current voltmeter with a high input impedance (say 10 megohms) is connected between ground rod X and ground D, a voltage (E_s) will be measured; this is the stray voltage, as shown in Figure 3.19b.

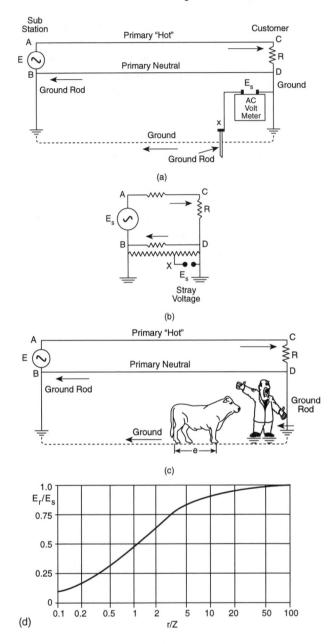

Figure 3.19. Origin of stray voltage (a), the equivalent circuit (b) and a subject receiving a mild stray-voltage shock (c). In (d) is shown how the source impedance (Z) of the equivalent stray-voltage generator can be determined by measuring the stray voltage (E_s) before placing a resistor (r) across the terminals of the voltmeter and after, to obtain the reading E_T.

To test the ability of the stray voltage source to drive current, a resistor r, (e.g. 500 ohms) is placed across the input terminals of the voltmeter and the reduction in voltage is noted while measuring the stray voltage. A large drop in voltage indicates the ability to deliver a low current. A small drop in voltage indicates the ability to deliver an appreciable current. The method of determining the source impedance of the equivalent stray-voltage generator will be described later. How much stray voltage and current are acceptable depends on the current pathway through a subject; there are those who believe that a few volts and a few milliamperes are excessive. Accounts of the various viewpoints are to be found in the Proceedings of the National Stray Voltage Symposium (1984).

Figure 3.19b presents the equivalent circuit, showing the stray voltage source (E_s). Note that in reality it is the current that strayed from the neutral that produced the stray voltage. The higher the current in the neutral, the higher the stray voltage. Likewise, the more distant the grounding rod X is from the customer ground at D, the more the stray voltage. The interesting fact is that if a subject (Ready Resistor) touches the customer ground at D and is standing on the ground, in good electrical communication with it, he will receive a shock due to the stray voltage, as shown in Figure 3.19c. Also shown in Figure 3.19c is Bess, a cow standing on the ground. Note that the stray current flowing in the ground will produce a voltage drop (e) between the hooves of the fore and hind limbs. If this stray voltage is large enough, Bess will receive a shock and will likely raise the hooves repeatedly, exhibiting a dancing appearance.

By placing a known resistor (r) across the input terminals of the high-impedance voltmeter used to measure the stray voltage (E_s), it is possible to obtain a good estimate of the (source) impedance (Z) of the equivalent stray-voltage generator. First, the open-circuit (no-resistor) stray voltage (E_s) is measured. Then the voltage is measured with the known resistance (r) placed across the terminals of the voltmeter; let this reading be E_r. By assuming that the impedances are resistive, by Ohm's law it is easy to show that the equivalent source impedance (Z) of the stray-voltage generator is $r(E_s/E_r - 1)$. By plotting (E_r/E_s) versus r/Z, as shown in Figure 3.19d, it is easy to determine Z.

EFFECT OF STRAY VOLTAGE

The stray current (I_s) that flows due to the stray voltage (E_s) is equal to the stray voltage divided by the impedance of the total circuit. In a practical case, this is constituted by the two contact impedances with the subject, the impedance of the subject, plus the equivalent impedance (Z) of the stray-voltage source. Stimulation occurs at the contact points with the subject where the current density is highest.

As stated previously, in some cases, animals have encountered shocks from stray voltage. Such shocks produce a behavioral rather than a physiological response. Animals are quick to learn to avoid an unpleasant stimulus such as a mild electric shock. Stray voltage in the milking area can cause cows to exhibit a constant stepping or dance-like behavior if enough stray voltage is applied to the hooves. A reluctance to eat or drink water may be the result of stray voltage shocks received when eating or drinking. When connected to a milking machine, a mild shock through the teats may result in uneven flow or cessation of milk delivery. Milk cows have even been reluctant to enter the milking area if such an act resulted in receiving a shock.

Table 3.3: Resistance of various electrical pathways through the cow

Pathway	Resistance		Current
Mouth to all hooves	350	324–393	60
	361	244–525[3]	60
Mouth to rear hooves	475	345–776[3]	60
Mouth to front hooves	624	420–851	60
Front leg to rear leg	300	250–405	60
	362	302–412	60
Front to rear hooves	734	496–1,152[3]	60
Rump to all hooves	680	420–1,220[3]	50
Chest to all hooves	980	700–1,230	50
	1,000	?	50
Teat to mouth	433	294–713[3]	60
Teat to all hooves	594	402–953	60
	880	640–1150	50
Teat to rear hooves	594	402–953[3]	60
Teat to front hooves	874	593–1,508	60
All teats to all hooves	1,320	860–1,960	50
	1,000	?	50
Udder to all hooves	1,700	650–3,000	60

˙From Lefcourt, A.M.

CURRENT PATHWAYS

Figure 3.20 illustrates a cow (Bess) connected to a milking machine, the housing of which is connected at M to the grounded neutral of the secondary of the two-phase (120–0–120 volt) transformer. The motor in the milking machine is connected between the hot (ungrounded) and neutral (P) sides of the transformer, the center tap of which (secondary neutral) is grounded by rod X. When the milking machine is operating, current flows in the hot and neutral conductors. Because the housing of the milking machine is connected to the neutral (at M), it will be at a potential (E_{PN}) above ground due to the voltage drop between points P and N of the neutral conductor. Note that this voltage (E_{PN}) will send current through the milk line, enter Bess via the teats (T), pass through Bess, and exit via the hooves and return via the ground to the grounding rod X, which is connected to point N, the secondary neutral. Thus, by standing on the ground, Bess completes the circuit. The sketch at the bottom of Figure 3.20 provides the equivalent circuit, showing that it is the voltage drop E_{PN} in the neutral conductor that sends current through Bess. Parenthetically it should be noted that E_{PN} will be larger if other appliances are connected to the transformer beyond the milking machine. This fact can be established by turning off all local equipment and measuring the stray voltage between a ground rod where Bess is standing and the local ground rod X.

Of the impedances shown in the equivalent circuit at the bottom of Figure 3.20, the highest impedances may be the ground path and the milk line. The resistance of the milk line can be calculated by knowing the resistivity of milk at body temperature, which was measured by the author as 170 ohm-cm at 36°C. For a 4-foot long, three-quarter inch diameter line filled with milk, the resistance is 7,400 ohms.

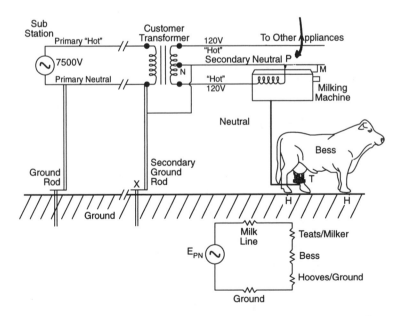

Figure 3.20. *Stray-voltage shock associated with a milking machine (top) and the equivalent electrical circuit (bottom).*

Although the foregoing example used the milk line to provide one contact point and the hooves the other, it is obvious that an animal standing on the ground and drinking water or eating food from a metal trough may receive a stray-voltage shock if the ground on which the animal is standing is at a different potential than that of the stanchion or trough. Also, it may be that other electrically-operated devices are mounted on environmental metal objects, resulting in a difference of potential existing between them and ground.

As stated previously, it is the stray voltage that causes current to flow. It is the current that produces the shock at the contact points. The current is equal to the stray voltage divided by the impedance of the entire circuit, a quantity that is often difficult to determine. However, an important part of that circuit is the animal. Lefcourt summarized the literature and his results are shown in Table 3.3; similar data for other animals are not available .

It is possible to eliminate the stray-voltage shock illustrated in Figure 3.20 by connecting the housing of the milking machine directly to the ground rod at X by a grounding conductor, (GC), as shown in Figure 3.21. In this way, irrespective of the voltage drop along the secondary conductors (hot, neutral, hot), the voltage between the housing of the milking machine and ground will be zero, providing there are no other appliances beyond Bess causing current to flow in the ground.

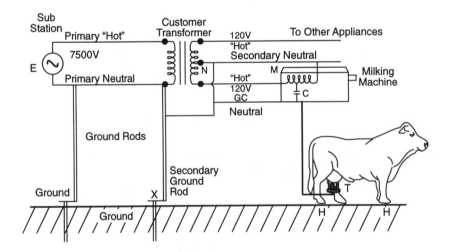

Figure 3.21. Elimination of the shock hazard by connecting the housing of the milking machine directly to the ground rod by a separate ground (GC) conductor.

Bonding, i.e. connecting together all of the exposed metal structures with heavy wire and connecting it to a ground rod, is the most common method of reducing the stray-voltage hazard. The animal support, e.g. ground or concrete, remains a stray-voltage hazard. The reinforcing metal in the concrete should be bonded to the main ground rod. Sometimes a metal screen or grate, bonded to the main ground, provides a way of eliminating hoof current. The various ways (mitigations) of dealing with a nonequipotential environment have been discussed by Gustafson (Agriculture Handbook).

An alternative method of guaranteeing that there is no teat current is to connect the housing of the milking machine to the bonded stanchion, food/drinking troughs and the support structure on which the animal is standing. In this way, all environmental metal surfaces will be at the same potential.

If in Figure 3.21, the grounding conductor (GC) becomes disconnected, a new hazardous situation is encountered. In this circumstance, the metal housing of the milking machine is capacitively coupled (C) to the hot side of the power line and leakage current will flow through Bess via the teats (T) to ground, which may result in a mild shock. In addition to capacitive leakage current, if there is moisture present in the motor, a resistive leakage path will be present and the leakage current will be larger. This subject was discussed earlier in this chapter in the section entitled "Leakage Current".

Those wishing to delve further into stray voltage will find a number of situations analyzed by Seevers in his book entitled "Ground Currents and the Myth of Stray Voltage". In this book it is pointed out that many electrically associated events are not due to stray voltage. His conversations with Elsie, the cow, are delightful reading.

THE 120/240 - VOLT SYSTEM

After about 1945, power receptacles appeared with three contacts to accept a plug with two blades and a rod. The two blades deliver the current to a device; the rod connects the metal housing of the device to ground. No power current is drawn through the grounding (green) conductor which is connected directly to the ground rod at the entry site of the power. Figure 3.22 illustrates such a 120/240-volt power-distribution system in which power is delivered at a high voltage (e.g. 7,500 volts) to a transformer at the customer site where the voltage is stepped down to 240 volts (120–0–120). The secondary center tap (neutral) is connected to ground (X) at the power-service entrance. The circular hole in the power receptacle is connected directly to the ground rod (X) by a green grounding conductor. Thus the housings of all metal-enclosed devices plugged into such receptacles will be at the same potential as the ground rod (X). This system is sometimes designated the 120/240-volt, three-wire system. Although the terms HOT, COLD and NEUTRAL are frequently used, the more correct terminology is UNGROUNDED and GROUNDED for the current-carrying conductors. The conductor (GC) that does not carry power current and connects the metal housing to ground at the service entrance is called the GROUNDING conductor.

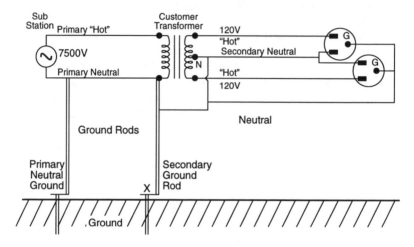

Figure 3.22. Conventional two-phase (120/240) power distribution system in which three-prong receptacles constitute the power out-lets at the customer site. The circular opening on the receptacle is connected directly to ground and is used to ground the metal housing of a device plugged into the receptacle. Power current is not drawn through this ground conductor (G).

STRAY-CAPACITIVE COUPLING

Two conductors separated by an insulator (dielectric) constitute a capacitor. Its capacitance depends directly on the area of the conducting surfaces and inversely on the separation between them. Alternating current can pass through a capacitor; the magnitude of the current depends on the capacitance, the voltage and the frequency. Thus, for a given frequency, the larger the area and the more closely spaced the conductors, and the higher the voltage, the greater the current.

Figure 3.23 illustrates a high-voltage transmission line, one conductor of which can be equated to one "plate" of a capacitor. A barbed-wire fence with wooden posts running parallel to the transmission line can be the other plate of the capacitor. If the transmission-line voltage is high enough, the fence is long enough and near enough, a person or animal standing on the ground and touching the fence could perceive current due to such stray capacitive coupling to the power line.

In Figure 3.23, below the transmission line is a wooden building with a metal roof; C_1 designates the capacitance between the metal roof and a hot (ungrounded) conductor of the transmission line; C_2 designates the capacitance between the metal roof and ground. A subject standing on the ground and touching the roof is capacitively coupled to the transmission line. In the equivalent circuit at the bottom of Figure 3.23, the subject is represented by the resistance R. On the right of Figure 3.23 is shown a truck parked below the transmission line. The metal body is capacitively (C) coupled to the transmission line and the tires provide some insulation from the ground. There is also a capacitance between the truck body and earth. A person standing on the ground and touching an exposed metal part of the truck can receive a mild shock. Measurements made by the author on automobiles parked below high-voltage transmission lines reveal voltages of 2 to 10 volts with respect to a rod driven into the ground.

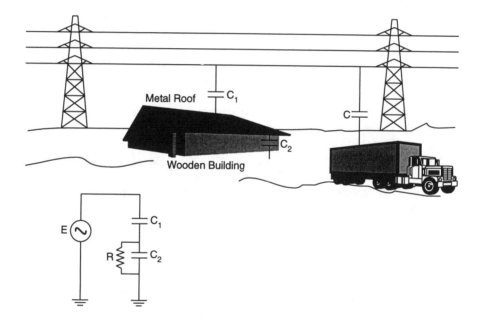

Figure 3.23. Capacitive coupling between a wooden building with a metal roof and the hot side of a transmission line. On the lower left is shown the equivalent circuit in which R is a subject standing on the ground and touching the metal roof. On the right is shown capacitive (C) coupling to a truck body.

In Figure 3.23, the shock hazard is eliminated if the metal roof is connected to ground by a grounding rod. Likewise the barbed wire fence on wooden fenceposts can be grounded in the same manner, thereby eliminating the shock hazard. Note in Figure 3.23, that there is a bird on one of the conductors. Because the bird is at the same potential as the conductor, it receives no shock. While it is true that the bird is capacitively coupled to an adjacent conductor of the transmission line, the capacitance is so small that no shock is perceived. The same is true for a lineman suspended in a basket from a single conductor of a transmission line.

ENVIRONMENTAL CONSIDERATIONS OF POWER LINES

From time to time the public media focus attention on the hazards associated with electrical power distribution systems, whether they be in the home or in the environment. Usually, epidemiological studies are cited to prove the existence of a hazard. It must be pointed out that epidemiological studies do not establish a causal relationship between two items or events; such studies only establish parallelism. For example, epidemiologically it can be noted that every morning after the cock crows, the temperature rises. This does not mean that it is the cock crowing that increases the temperature. As everyone knows, the cock crows at sunrise and it is the rising sun that increases the temperature. Therefore when viewing epidemiological data, it is useful to ask if the parallelism may be due to some other agent.

Three methods are used to investigate the relationship of one event with another; they are: 1) epidemiological, 2) in vitro and 3) in vivo. The nature of epidemiological studies has been described in the preceding paragraph. In-vitro studies involve quantitative measurement of the effect of the suspected agent (by itself) on cells in a culture medium. In-vivo studies involve making quantitative measurements on animals and man of the effect of the suspected agent only. In both cases, the goal is to obtain what is called a dose-response curve, which is a graph showing a quantitated response versus a quantitative increase in the intensity or quantity of only the suspected agent.

Although many studies have been published on the possible effects of high-voltage transmission lines on animals and man, only three will be reviewed here; the first by Amstutz (1980) dealt with a 765-kV, 60-Hz transmission line running through farms where animals were raised; the second is in the New York State Power Lines Project report, which also deals with a 765-kV, 60-Hz line, and includes epidemiological studies. The third study by Raleigh (1988) involves raising animals, alfalfa and wheat under a 500-kV, direct-current line.

When investigating the effects of high-voltage transmission lines, three relevant parameters can be measured: 1) the electric field (kilovolts-/meter), 2) the magnetic field in Gauss and 3) the ionization products produced in the environmental air. It is also useful to remember that all living things on earth evolved in the presence of the static earth's magnetic field which ranges from 0.5 Gauss over the north magnetic pole to 0.79 Gauss over the south magnetic pole. The intensity near the equator is 0.3 to 0.4 Gauss.

Amstutz and Miller (1980) measured the electric and magnetic fields under a 765-kV line running through Indiana and Michigan. Farms with animals near and under this line were selected for study; three raised dairy cattle, one raised sheep, four raised beef cattle, one raised swine and beef cattle and one raised horses and mixed cattle. In all, the study involved 55 sheep, 337 hogs and 429 dairy cattle. The electric field strengths under the line ranged from 2.5–12 kV/meter and the magnetic field strengths ranged from 4 to 56 milligauss. Currents were measured to ground in humans standing under the line; the currents ranged from 5 to 140 microamps. Questionnaires were filled out bimonthly by the farmers, and Amstutz and Miller made bimonthly visits to the farms. At the end of the project, interviews with the farmers were conducted by two different investigators (Sunier and Roy). The report concludes as follows: "At the end of this two-year study, nine of the original eleven participants still remained. Two dropped out in about the middle of the program, one due to personal health problems, the other terminated livestock farming operations. The remaining nine farm operators were individually interviewed by R.E. Sunier of the I&ME Co. and W.R. Roy of AEPSC concerning their opinion of this study of 765-kV transmission lines, and especially any 765-kV effects on farm animals. The interviews were open to any questions they might have concerning 765-Kv transmission. The following represents the composite views of these nine farm participants.

"The 765-kV lines have no adverse effect on farm animals. Some of the group even felt that the animals preferred to be around 765-kV towers and under the line as compared to open field. Since this study covered farms raising sheep, dairy cattle, beef cattle, horses and hogs, it presents a wide spectrum of observations on many different species of farm animals. No farmer in this study has noted behavioral changes to any of his animals that he could attribute to 765-kV electrical effects.

"All of the farmers would prefer, however, that the line was not on their property. The basic reason given is the nuisance it causes when farming around towers or guy lines, especially with large farm equipment. It requires more careful operation of farm equipment in these areas. All of the farmers accept the fact that electrical energy transmission is an unavoidable national necessity and, consequently, transmission lines are acceptable as long as only nuisance factors are involved. Some of the farmers noted nuisance shocks from insulated objects such as vehicles parked underneath the circuit or metal roofs on otherwise non-metallic buildings. Most of them did not consider this problem significant enough to report to the power companies prior to the interview. Some mentioned the fact that nuisance shocks from weeds could be received on the ankles, especially when walking under the line with rubber-soled shoes. None of the farmers felt this was a serious consideration with the line.

"All participants felt that it was beneficial and proper for an electric company to sponsor this type of study in order to confirm more positively what the farmers themselves can observe; that is, that there are no effects on livestock from the lines. Three of the remaining nine participants purchased their farms after the 765-kV transmission lines were already on the property. One farmer was leasing a farm with the transmission line on it. None of these four farmers were negatively influenced by the fact that the line was on the property when they contemplated farming operations at these locations. Their subsequent experience has confirmed the fact that the line is not a deterrent to successful animal husbandry."

The New York State Power Lines Project summary states that the "Project was established to conduct research and to review the scientific literature to determine whether health hazards of these fields are possible. Particular attention was directed to the fields generated by 765-kV overhead transmission lines. The research program, supported by contributions assessed from all New York State electric utilities, provided support to 16 different research groups studying human, animal and isolated cell sensitivity to electric and magnetic fields. Most of the research studies reported no effects of concern. Of the few effects, some warrant further consideration.

"No effects were found on reproduction, growth or development. Several studies showed no evidence of genetic or chromosomal damage that might lead to inherited effects or cause cancer. While most measurements of behavior and brain function did not demonstrate changes, some did show changes that were small but consistent. Some of these appear to result from changes in body rhythms, and might interfere with normal sleep patterns. There were also changes in pain responses and in the ability of rats to learn."

The childhood cancer studies were epidemiological and the Project recognized this fact and stated: "Because Wertheimer and Leeper (1979; 1982) reported an association between residential exposure to magnetic fields and incidence of cancer in children and adults in Denver, Colorado, two epidemiological investigations were done to a) study incidence of cancers in children in the Denver, Colorado area (Savitz, Appendix 15) and b) study the incidence of adult nonlymphocytic leukemia in Seattle, Washington (Stevens, Appendix 16) as a function of residential exposure to electric and magnetic fields. [The references and Appendices can be found in the Report.]

"In the Savitz case-control study, all cases of childhood cancer (ages 3–14) between 1978 and 1983 were selected. Controls were identified through random digit dialing and matched by age and sex. Exposure was estimated by determination both of wiring configuration outside the home, as developed by Wertheimer and Leeper (1979), and by direct measurement of the fields. Wiring configuration was found to correlate with the field measurements, and the major factor contributing to magnetic fields was found to be distribution lines. There was a positive association between wiring configuration and increased cancer risk. This held for all childhood cancers, especially for leukemias, and to a lesser degree, for brain tumors. There appeared to be a dose-response relation, in spite of the inexact measures of exposure. The relative risk was above 2 for the highest exposed group. No sources of bias were identified to explain the results, although the somewhat limited response rate remains a concern.

"In the study of adult non-lymphocytic leukemia (Stevens) similar procedures in general were used but no association between cancer and magnetic field exposure, as measured by wiring configuration, direct field measurement or an engineering-based code, was found in the 164 cases and 204 controls."

As stated earlier, epidemiological studies do not link cause and effect; they alert the investigator to identify all possible factors that could link two events.

The Raleigh study (1988) employed a 500-kV, direct-current line. With a constant current in such a line, only a static magnetic field would be produced, the polarity of which would add to or subtract from the earth's magnetic field. However such a line does not carry a static current; it varies from time to time as the load changes. The current in the line studied (Cello-Salmar) ranged from 500 to 2,000 amps. The 3-year study involved 10 power companies in the US and Canada. The electric field (kV/m), magnetic field (mG), ion current density (mA/M^2) and ion density (ions/cm^3) were measured under the line and at several distances from the center of the line. In addition a considerable amount of meteorological data were acquired.

The Raleigh study was conducted in central Oregon. One hundred cows and 6 bulls were contained in a large pen under the line. An equal control group occupied a pen 2,000 feet west of the line. The management techniques were identical for both groups. Food and water intake, reproduction and behavior were all observed.

The summary stated "There were no significant differences in feed, mineral or water consumption between line and control herds. One of the line herds showed a preference for water fountains located beneath the negative conductor, but the other line herd failed to show the same preference. Since the fountain preference was not repeated, there is no apparent link with the power line. An intensive disease prevention program helped keep the animals healthy. Relatively few health problems occurred during the study, and the diseases which did occur were almost equally prevalent in both line and control groups.

"In this study, six control cows developed cancer eye (squamous cell carcinoma), four line cows had cancer eye and one had adenocarcinoma at slaughter. Cancer eye appears to be genetic, most common in Hereford and Hereford cross cattle, and occurs more frequently as the cows age. The study herds were made up primarily of Hereford and Hereford-cross cattle, typical of commercial beef herds.

"Mortality was also very low. A total of 20 animals died during the study. Two cows, one bull, and seven calves died in the line herd, while three cows and seven calves died in the control herd. This represents a mortality rate of 2.5 percent. A typical mortality rate for range livestock operations is about 5.4 percent.

"The three breedings that occurred during the study resulted in an average conception rate of 94.3 percent. The [line] animals had a conception rate of 94.7 percent, while the controls had a conception rate of 94 percent. There was no significant difference in conception between the line and control groups.

"There was no significant difference in the number of calves weaned between the line and control herds. The line [animals] had a weaning percentage of 94 percent and the control was 92 percent. There was no significant difference between the date of birth between groups with the average being the 90th Julian day. There was no significant difference in calving interval for the two intervals between groups with mean intervals of 411 days for 1985–86 and 347.4 days for 1986–87.

"Average daily gain (ADG) did not differ between line and control treatment groups. However, the 1985 calf crop had significantly lower ADG than the other 2 years. This was attributed to several factors. Mainly, the cows were not in good condition when they first arrived at the facility; some were young and still growing while nursing their calves and, as range animals, were not accustomed to pen confinement.

"Adjusted weaning weights (AWW) did not differ between line and control treatments. However, there was a difference between years; the 1985 calf crop had significantly lower AWW than the other two calf crops. The difference was attributed to the same factors that affected the ADG in 1985.

"Sexual development of the bull calves was evaluated at weaning. There were a few calves in both groups which were considered to have abnormal testicles. However, the rates of variation were within normal levels and did not differ significantly between groups.

"Cow weights at weaning did not significantly differ between groups. However, the cows in 1985 weighed significantly less than they did in 1986 or 1987. This difference can be attributed to the same factors which caused the smallest ADG and AWW in 1985.

"At the end of the study the cows were rated for condition and then slaughtered. There was no significant difference in condition ratings between the groups. Carcass weights were obtained at slaughter, and there was no significant difference in carcass weights between groups.

Based on clinical manifestations and pathological findings, the d-c power line had no demonstrable influence on the cattle or their offspring.

"Monitored aspects of livestock behavior included: arrangement of cattle at feedbunks, afternoon resting or loafing distributions, night time bed site distributions, 24-hour distribution patterns, and cattle activities for 24-hour periods. A small but significant ($P<0.05$) difference in positioning of the average cow at the feedbunk was detected in group A with the disparity averaging only 2.9 feet. As this is slightly less than the width of a mature cow and the average treatment and control cow fed within 1.7 feet of center line, the difference was judged to be of no biological significance.

"Seven out of eight comparisons of livestock distribution patterns during afternoon, night, and 24-hour periods detected significant ($P<0.05$) line: control differences. When differences were examined for evidence of some pattern relative to the position of HVDC conductors, small but statistically significant relationships were found. This suggests 1 to 4 percent of the line cattle were emigrating to areas 100 to 150 feet from the D-C line. This occurred in 12 of 54 attempts to fit models to the data (mean r^2 of significant models = 0.56). The majority of cattle still focused their activities in cells either under or adjacent to the line. Although adjustments in management are not deemed necessary if cattle are maintained under a D-C line, allowance of loafing or resting areas 100 to 150 feet from center line would suffice if one wished to accommodate the responses expressed by a small proportion of the cattle.

"Efforts to correlate positioning of line cattle at feedbunks or presence of line cattle in the pen's central cells to measures of electric fields or audible noise were unsuccessful. Given the line/control distribution disparities, this would suggest that cattle: 1. developed patterns or habits in behavior at the onset of the study, with no subsequent responses to variation of electrical effects, 2. behavior of the majority of the cattle masked responses exhibited by a small proportion of sensitive animals. 3. cattle were insensitive to analyzed variables, or 4. line/control disparities were related to structural differences between line and control areas.

"Significant differences ($P<0.05$) were detected in group A cows and calves for drinking and nursing activities. Differences were less than 1 percent and seemed of no biological significance. No differences were found in the major activities of bedding, eating, standing or walking.

"The exposure levels to electric fields and air ions for the line pens and control pens were evaluated. The average field and air ion levels in the control area were only a few times the observed levels with the line off. When the wind was blowing away from the control area toward the line, the control area should have been at ambient levels."

REFERENCES

Alexander, L. Electrical injuries to the central nervous system. Med. Clin. N. Am. 1938, 22:663–688.

American Association for the Advancement of Medical Instrumentation (AAMI) Safe Current Limits for Electromedical Apparatus SCL 12/78 (1978). AAMI, Arlington, Virginia.

Amstutz, H.E., and Miller, D.B. A Study of Farm Animals Near 765-kV Transmission Lines. Second Printing 8/80. Indiana & Michigan Electric Co. and the American Electric Power Service Corp. 31 pp.

Appleman, R.D. In Effects of Electrical Voltage/Current on Farm Animals. US Dept. Agriculture, Agriculture Handbook No. 696. (No date).

Arieff, A.J. Threshold studies in electrical convulsions using a square wave stimulatory. Quart. Bull. Northwestern U. Medical School 1948,22:10–16.

Bayles, S. Square waves (BST) versus sine waves in electroconvulsive therapy. Amer. J. Psychiat. 1950,107:34–41.

Beck, C.S., Pritchard, W.H. and Feil, H.S. Ventricular fibrillation of long duration abolished by electric shock. JAMA 1947,135:985–986.

Bernstein, T. A grand success. IEEE Spectrum 1973, 10:54–58.

Bidder, T.C., Strain, J.J. and Brinschwig, L. Bilateral and unilateral ECT. Amer. J. Psychiat. 1970,127:739–745.

Blair,H.A. 1932. On the intensity-time relations for stimulation by electric currents. J. Gen. Physiol. 15:177–185, 709–729,731–755.

Cerletti, U. Old and new information about electroshock. Amer. J. Psychiat. 1950, 107:87–94.

Canadian Standard Association, 235 Montreal Road, Ottawa 7, Canada. Code for the Use of Flammable Anesthetics. Canadian Standard Z 32, 1963.

Cronholm, B. and Ottorson, J.O. Ultrabrief stimulus technique in electroconvulsive therapy. J. Nerv. Ment. Dis. 1963,137:117–127.

Dalziel, C.F. Effects of electric shock on man. IRE Trans. Med. Electron. 1956,-5:44–62.

Denier, A. L'electronarcose. Anesth. et Analg. 1938,4:451–471.

Emerson, S.J. Quoted by Lee, W.R., 1961.

Ferris, L.P., King, B.G, Spence, P.W. and Williams, H.B. Effect of electric shock on the heart. Elect. Eng. 1936, 85:498–515.

Friedman, E. Unidirectional electrostimulated convulsive therapy. Amer. J. Psychiat. 1942–43, 99:218–223.

Frostig, J.P., Van Harreveld, A., Reznick, S., Tyler, D.B. and Wiersma, C.A.G. Electronarcosis in animals and man. Arch. Neurol. Psychiat. 1944,51:232--242.

Geddes, L.A. Electronarcosis. Med. Elect. and Biol. Engng. 1965, 3(1):11–26.

Geddes, L.A., Hoff, H.E. and Voss, C. Cardiovascular-respiratory studies during electronarcosis in the dog. Cardiovascular Research Center Bulletin, 1964, 3(2):38–47.

Geddes, L.A. and Hamlin, R. The first human heart defibrillation and the defibrillator. Amer. J. Cardiol. 1983,52:403–405.

Geddes, L.A., and L.E. Baker. 1971. Response to the passage of electric current through the body. J. Assoc. Adv. Med. Instrum. 5:13–18.

Gustafson, R. Mitigation In Effects of Electrical Voltage/Current on Farm Animals. U.S. Dept. Agriculture. Agriculture Handbook No. 696.

Hardy, J.D., Turner, M.D. and McNeil, C.D. Electrical anesthesia. J. Surg. Res. 1961, 1:152–168.

Hardy, J.D., Carter, T. and Turner, D. Catecholamine metabolism. Ann. Surg., 1959, 150:666–683.

Hardy, J.D., Fabian, L.W. and Turner, M.D. Electrical anesthesia for major surgery. J. Amer. Med. Ass., 1961, 175:599–600.

Hassin, G.B. Changes in the brain in legal electrocution. Arch. Neurol. Psychiat. 1933, 30:1046–1060.

Hayes, K.J. The current path in electrode convulsive shock. Arch. Neurol. Psychiat. 1950,63:102–109.

Hoff, H.E. Nutrition of the Heart. In A Textbook of Physiology. J.F. Fulton. 1955 W.B. Saunders.

Hooker, D.R., Kouwenhoven, W.B. and Langworthy, O.R. The effect of alternating electrical current in the heart. Amer. J. Physiol. 1933, 103:444–454.

Jaffe, R.H. Electropathology. Arch. Pathol. 1928, 5:837–870.

Jex-Blake, A.J. Death by electric current and lightning. Brit. Med. J. 1913, March 1:425–430.

Knutson, R.C. Experiments in electronarcosis. Anesthesiology 1954,15:551–558.

Knutson, R.C., Tichy, F.Y. and Reitman, J.A. The use of electrical current as an anesthetic agent. Anesthesiology 1956, 17:815–825.

Kouwenhoven, N.B. and D.R. Hooker. Electric shock; effects of frequency. Elect. Eng. 1936, 55:384–386.

Langworthy, O.R. Abnormalities in the central nervous system by electrical injuries. J. Exp. Med. 1930, 51:943–968.

Lee, W.R., and J.R. Scott. Thresholds of fibrillating leakage currents along ventricular catheters. Cardiovasc. Res. 1973, 7:495–500.

Lee, W.R. A clinical study of electrical accidents. Brit. J. Ind. Med. 1961, 18: 260–269.

Lefcourt, A.M. In Effects of Electrical Voltage/Current on Farm Animals. US Dept. Agriculture Agriculture Handbook. No. 696. (No date).

Liberson, W.T. Brief stimulus therapy. Amer. J. Psychiat. 1948, 105:28–39.

Lorimer, R., Segal, M. and Stein S. Path of current distribution in brain during electroconvulsive therapy. EEG Clin. Neurophysiol. 1949, 1:343–348.

MacDonald, C.F. The infliction of the death penalty by means of electricity. Trans. Med. Soc. State N.Y. 1892,400–427.

Monsees, L.R., and D.G. McQuarrie. Is an intravascular catheter a conductor? Med. Electron. Data 1971, 12:26–27.

National Stray Voltage Symposium. Oct 10–12, 1984, Syracuse, NY. Amer. Soc. Agricultural Engineers. 2950 Niles Rd., St. Joseph, MI 49085-9659.

N.F.P.A. National Fire Protection Association, 470 Atlantic Avenue, Boston, MA 07210, National Electrical Code (NFPA #70), Essential Electrical Systems for Hospitals (NFPA #76), Flammable Anesthetics Code (NFPA #56), Inhalation Therapy, 1968 (NFPA #56B), Nonflammable Medical Gas Systems, 1967 (NFPA #565), Bulk Oxygen Systems at Consumer Sites, 1965 (NFPA #566), Static Electricity, Recommended Practice on, 1966 (NFPA #77)

New York State Power Lines Project, Final Report. Biological Effects of Power Line Fields. DOE/ER/60099–T3 and DE 88–002346. 154 pp & Appendices.

Prevost, J.L. and Batelli, F. Death by electric current (alternating current). Comptes Rendus Acad. Sci. (Paris) 1899, 128:668–670.

Pritchard, E.A. Changes in the nervous system due to electrocution. Lancet 1934, 1:1163–1167.

Raleigh, R.J. University of Oregon, Joint HVDC Agricultural Study; Final Project Report, Sept 30, 1988. Bonneville Power Administration, Portland OR. Contract #DE-A179-85BP21216. 289 pp plus figures and appendices.

Roy, O.Z., J.R. Scott, and G.C. Park. 60 Hz ventricular fibrillation and pump failure threshold versus electrode area. IEEE Trans. Biomed. Eng. 1976, BME-23(1):45–48.

Roy, O.Z., G.C. Park, and J.R. Scott. Intracardiac fibrillation threshold as a function of the duration of 60Hz current and electrode area. IEEE Trans. Biomed. Eng. 1977, BME-24(2):430–435.

Sances, A.D., Szablya, J.F., Morgan, J.D. et al. High-voltage powerline injury studies. IEEE Trans. Power Appar. System 1981, PAS100(2):552–558.

Sances, A.D., Myklebust, J.B., Larson, S.J. et al. Experimental electrical injury studies. J. Trauma 1981, 21(8):589–597.

Schneck, J.M. The history of electrotherapy. Amer J Psychiat. 1959, 116:463–464.

Seevers, O.C. Ground Currents and the Myth of Stray Voltage. Liburn, GA, 1988 Fairmont Press 209 pp.

Smitt, J.W. and Wegner, C.F. On electric convulsive therapy with particular regard to parietal application of electrodes controlled by intracerebral voltage measurements. Acta. Psychiat. Scand. 1944, 19:529–549.

Spitzka, E.A. and H.E. Radash. The brain lesions produced by electricity as observed after legal electrocution. Am. J. Med. Sci. 1912, 144:341–347.

Staewen, W.A., M. Mower, and B. Tabatznik. The significance of leakage currents in hospital electrical devices. J. Mt. Sinai Hosp. (N.Y.) 1969, 15:3–10.

Starmer, C.F., and R.E. Whalen. Current density and electrically induced ventricular fibrillation. Med. Instrum. 1973, 7(2):158–161.

Valentine, M., Keddie, M.G. and Dunne, D. A comparison of techniques of electroconvulsive therapy. Brit. J. Psychiat. 1968, 114:980–996.

Van Harreveld, A., Tyler, D.B. and Wiersma, C.A. Brain metabolism during electronarcosis. Amer. J. Physiol. 1943, 139:171–177.

Van Harreveld, A. On the mechanism and localization of the symptoms of electro-shock and electronarcosis. J. Neuropath. exp. Neurol. 1947, 6:177–184.

Van Harreveld, A. and Dandiliker, W.B. Blood pressure changes during electro-narcosis. Proc. Soc. exp. Biol. (N.Y.) 1945, 60:391–394.

Van Harreveld, A. and Tietz, E.B. Effect of electronarcosis on level of "adrenaline-like" compounds in blood. Proc. Soc. exp. Biol. (N.Y.) 1949, 70:496–498.

Weaver, R.W. and Rush, S. Current density in bilateral and unilateral ECT. Biol. Psychiat. 1976, 11:103–112.

Weinberg, D.I., and J.C. Artley. Electric shock hazards in cardiac catheterization. Circ. Res. 1962, 1:1004–1009.

Whalen, R.E., C.F. Starmer, and H.D. McIntosh. Electrical hazards associated with cardiac pacemaking. Ann. N.Y. Acad. Sci. 1964, 111:922.

Wilcox, P.H. Electroshock therapy. Amer. J. Psychiat. 1947, 104:100–112.

Wilcox, P.H. Brain facilitation and brain destruction: the aim in electroshock therapy. Dis. Nerv. System 1946, 7:201.

Wulfsohn, N.L. and McBride, W. Electronic anesthesia. S. Afr. Med. J. 1962, 36:941–943.

Zoll, P.M., Linenthal, A.J., Gibson, W. et al. Termination of ventricular fibrillation in man by externally applied countershock. New Engl. J. Med. 1956, 254:727–732.

Chapter 4
LIGHTNING INJURIES

INTRODUCTION

Lightning is a ubiquitous natural phenomenon with the potential for producing many different types of injury. Jex-Blake (1913), in the third Goulstonian Lecture presented to the Royal College of Physicians (London), gave the first statistics on death by lightning; table 4.1 summarizes his findings. He pointed out that the death rate in cities was much less than in rural areas. Viemeister (1972) estimated that there are 1,800 thunderstorms occurring at any given moment throughout the world, with 44,000 thunderstorms being born each day, producing 8,000,000 flashes. He estimated that 180 people are killed by lightning annually in the US; but stated that many who are struck are stunned and recover. More recent data indicate that Viemeister's death-rate estimate is a little high. Cox (1992) reported that the number of deaths from lightning in the UK between 1984 and 1989 was 15, i.e. about 3 per year. According to Cox (1992), Cooper estimated that there are about 150–200 lightning deaths per year in the US (i.e. about one per million people) and probably about ten times the number of injuries. Ohashi (1986) reported that in Japan in the 17 years before 1975, there were 140 lightning accidents, of which 50 sustained damage to their clothing, including skin burns, and stated that there was better than 50% survival in patients with such symptoms. In Sweden, Eriksson (1988) reported that the death rate from lightning was 0.2 to 0.8 per million people. To understand the types of injury, it is useful to know the nature of lightning.

NATURE OF LIGHTNING

Our first scientific knowledge of the nature of lightning came from Benjamin Franklin who conducted his many well-known experiments on electricity, beginning about 1750. Before that time, it was known that certain substances, e.g. glass, wood, sulfur balls etc., when rubbed with silk or wool, became electrified and attracted or repelled objects and often produced long sparks if rubbed briskly. At that time it was believed that there were two kinds of electricity, artificial and natural (lightning).

In 1747 the Leyden jar (capacitor), capable of storing substantial charge (electric fluid) was described by von Kleist of Karmin and about the same time by van Musschenbroek of Leyden. With the Leyden jar, the charge produced by frictional electricity could be stored and stout sparks were produced when the jar was discharged.

Table 4.1: First statistics on lightning death rate
per million per annum.*

Country	Rate
UK (1852–1880),	0.88
UK (1901–1910)	0.36
Europe	0.88
Hungary	16
Syria - Corinth	10
Prussia	4.4
France	3
Sweden	3
Belgium	2
U.S.	10

* From Jex-Blake, Brit. Med. Journ. 1913

From his experiments, Franklin introduced two new concepts: 1) resting bodies were neutral; friction caused them to be charged, and 2) a pointed metal rod collected charge and was also effective in giving it off with the production of a spark. Franklin believed that thunderclouds were charged (electrified) and proved this belief with tall pointed conductors and his famous kite experiments which drew charge down the kite string from the passing thunderclouds, the evidence for which was a spark at the lower end of the string which, when wet, was conductive. The charge brought down the kite string was used to charge a Leyden jar which could be dishcarged, providing a stout spark. It was from these experiments that it was shown that natural electricity (lightning) and artificial (frictional) electricity were the same.

Lightning is defined as a transient, high-current electrical discharge that occurs when an area of the atmosphere attains an electric charge sufficient to produce an electric field (volts/meter) strong enough to break down the insulation provided by the intervening air. Such a discharge may travel from a charged cloud to ground, from the ground to a charged cloud, between charged clouds, or within a charged cloud. Charge is produced in a cloud in which there is cold dense air aloft and warm moist air below. The warm air rises in a strong updraft and the cold air descends in a strong downdraft to form a thundercloud; these strong air currents constitute a cell. In the cloud there is a temperature gradient and a turmoil of wind, water and ice crystals, as described by Malan (1989) and shown schematically in Figure 4.1.

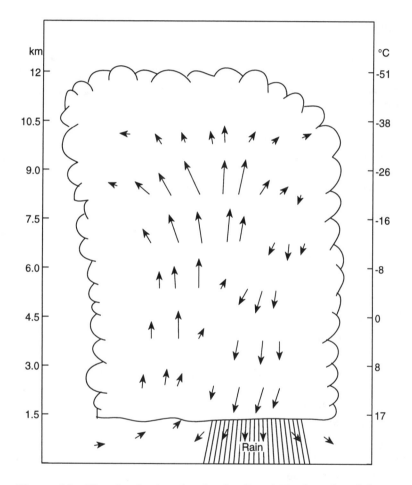

Figure 4.1. Sketch of a thundercloud cell with the lengths of the arrows representing wind velocity. The left ordinate identifies altitude (kM) and the right identifies temperature (°C) (From Malan 1989).

The natural history of a cell has three successive stages: 1) cumulus, 2) mature and 3) dissipative. In the cumulus stage there is a strong updraft near the top of the cell. The temperature in the cell is higher than that of the air surrounding the cloud at the corresponding height. No precipitation occurs in this stage. The mature stage starts when the water droplets and ice crystals can no longer be supported by the updraft and rain starts to fall. Figure 4.1 shows a thundercloud in its mature stage. In the dissipative stage the downdraft becomes larger, spreading throughout the cell and the rate of rainfall diminishes, becoming a steady shower. According to Malan (1969) a typical isolated towering cumulo-nimbus cloud extends to about 12 km and has a diameter of 8 km, the diameter of the cell being about 1.5 km. Note that, in Figure 4.1, the temperature at the top of the cloud is –50°C.

Table 4.2: Components of a lightning flash and return stroke[*]

	Minimum	Representative	Maximum
Stepped leader			
Length of step, m	3	50	200
Time interval between steps, μsec	30	50	125
Average velocity of propagation of stepped leader, m/sec	1×10^5	1.5×10^5	2.6×10^5
Charge deposited on stepped-leader channel, coul	3	5	20
Dart leader			
Velocity of propagation, m/sec	1.0×10^6	2.0×10^6	2.1×10^6
Charge deposited on dart-leader channel, coul	0.2	1	6
Return stroke			
Velocity of propagation, m/sec	2.0×10^7	5.0×10^7	1.4×10^8
Current rate of increase, kA/μsec	<1	10	>10
Time to peak current, μsec	<1	2	30
Peak current, kA		10–20	110
Time to half of peak current, μsec	10	40	250
Charge transferred, excluding continuing current, coul	0.2	2.5	20
Channel length, km	2	5	14
Lightning flash			
Number of strokes per flash	1	3–4	26
Time interval between strokes in absence of continuing current, msec	3	40	100
Time duration of flash, sec	10^{-2}	0.2	2
Charge transferred including continuing current, coul	3	25	90

[*] from Uman 1969

The upper part of a thundercloud is preponderantly positively charged and the lower part of the cloud is mainly negatively charged, which induces a positive charge on the ground. As the charge buildup increases, the dielectric breakdown of the intervening air is approached. There then occurs a localized branched breakdown producing a stepped leader extending a short distance from the cloud and reducing the negative charge. Very shortly thereafter (in microseconds) another stepped leader develops extending earthward. This stepping process continues and increases the electric field at the earth's surface and produces upward-moving leaders. When a downward-going leader encounters an upward-going leader, usually originating from an object above the ground, a bright lightning strike occurs.

A lightning flash is seen to flicker. Shutterless high-speed photography of flashes reveals that the discharge progresses in steps, which results in a cloud-to-ground stroke.

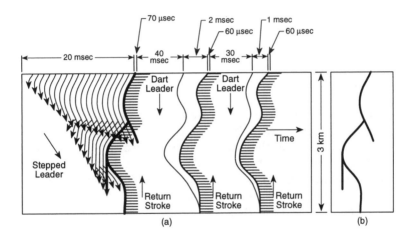

Figure 4.2. Temporal aspects of the development of a lightning flash from a height of 3kM showing the stepped leader and the return stroke. (From Uman 1969 by permission of the author).

Figure 4.2 shows the temporal aspects of the stepped leader and the return stroke, along with the type of flash seen by the eye (right). Table 4.2 presents information on the components of a lightning flash which consists of many short-duration (μsec) pulses. The current in a single stroke is many tens of kiloamperes. The peak temperature associated with a stroke is on the order of 20,000 to 30,000°C. The color of the flash represents ionization of the atmospheric gas molecules and the plasma produced by the high electric field. Malan (1969) estimated that the energy expended in a single flash is 10^{10} joules and that the potential between a cloud and ground is on the order of 10^8 to 10^9 volts. The sudden liberation of energy produces an acoustic shock wave, an intense temperature pulse and electromagnetic radiation (ultraviolet, visible and radio frequency). The frequency of the current pulses in the flash is about 25/sec. Obviously there are many ways that a lightning flash can cause injury.

PATHOPHYSIOLOGICAL RESPONSES

The types of pathophysiological responses due to a lightning strike depend on the nature and intensity of the current and the pathway of the current through the body which can be modeled as an electrolytic resistor enveloped by an insulator (skin). However, with a lightning strike, the skin dielectric is broken down due to the very high voltage gradient associated with lightning. The intense liberation of energy produces a shock wave that can tear off clothing. In addition the intense heat can burn the clothing and skin. Moreover, the bright flash of light can cause eye damage and the acoustic shock wave can damage the ear. With these facts in view it is appropriate to examine some of the ways that lightning can interact with the body.

STEP VOLTAGE

When lightning strikes the ground, current flows therein and a sudden voltage rise appears between two points on the ground, the potential being known as the step voltage or earth-potential rise (EPR); Figure 4.3 illustrates the concept. The magnitude of the voltage (V) depends on the spacing (s) between the two points, the current (I), the resistivity (ρ) of the earth and the distance (d) to the lightning strike. Andrews et al. (1992) provided the following expression for the step voltage (V):

$$V = \frac{I\varrho}{2\pi}\left[\frac{s}{d(d+s)}\right] \qquad (1)$$

For a subject standing on the ground, the greater the spacing (s) between the feet, the larger the step voltage. For large four-footed animals, the step voltage will be higher than for man. The current due to the step voltage will flow through the legs and the lower torso, with little reaching the heart and even less reaching the medullary respiratory center at the base of the brain. Because the lightning current is a series of very short-duration (μsec) pulses, the likelihood of stimulating the heart is small owing to its fundamental excitability characteristics (see Chapter 2). However, the subject could perceive a sensation and perhaps a few muscle contractions. If the step voltage is high enough, animals may be startled and may bolt. Shoes or boots with nails in the soles or bare feet on wet ground both favor a high step current through the limbs.

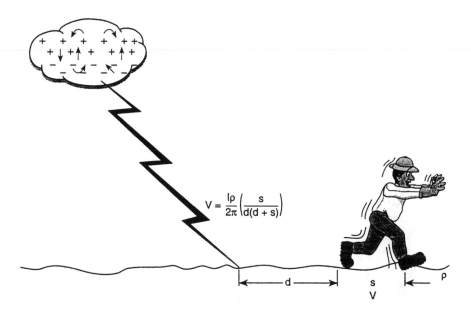

$$V = \frac{I\rho}{2\pi}\left(\frac{s}{d(d + s)}\right)$$

Figure 4.3. Step voltage (V), the sudden rise in potential between two points on the ground a distance s apart and at a distance d from the lightning strike.

LIGHTNING STRIKE

Obviously a direct strike by lightning can produce several types of injury. However, before discussing them, the mechanisms that cause current to flow through the body will be described. Although the events associated with a lightning strike are complex, a simple model can serve to provide enlightenment. For example, the body can be modeled as an elongated electrolytic resistor (the tissues), surrounded by a reasonably good insulator or dielectric, (the skin), as sketched in Figure 4.4A. Typically the resistivity (ρ) of body fluids ranges from about 50 to 200 ohm-cm and organs have a resistivity ranging from 300 to 3,000 ohm-cm, that for bone being much higher. The electrical equivalent circuit is shown in Figure 4.4B.

Just before the lightning flash, the stepped leader will cause short-duration (μsec) pulses of current to flow through the skin capacitance which has a low impedance for such rapidly rising, short-duration pulses and the current (I) that flows is sketched in Figure 4.4C. When the electric field becomes high enough, the skin dielectric breaks down and the current flow through the subject will increase suddenly, as shown in

Figure 4.4D. A typical resistance for the body is 1,000 ohms and the current for a lightning strike is about 20,000 amps; therefore the head-to-foot voltage will be 20,000 × 1,000 = 20 million volts. The dielectric breakdown for air is about 75,000 volts/in. Assuming a 6 foot (72") tall subject, the air breakdown voltage for 72 inches is 72 × 75,000 = 5.4 million volts. Note that the head-to-foot voltage is about four times this value and therefore an arc passes over the surface of the body to ground, thereby providing a parallel current path and reducing the current flow through the body, as shown in Figure 4.4E. The foregoing sequence of events has not been documented, but is predicted on the basis of the nature of lightning, the composition of living tissues and the types of injury.

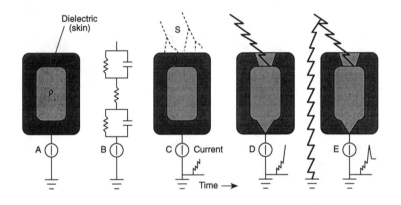

Figure 4.4. Simplified model for the body (A) consisting of an electrolytic conductor (ρ) surrounded by a dielectric (skin). In B is shown a simple equivalent circuit. The approach of a stepped leader (S) is shown in C, causing pulses of current (I) to flow through the body via its dielectric envelope. In E, the dielectric breaks down and the body current rises, then falls when a flashover occurs.

Figure 4.5A illustrates a subject who is struck by lightning. As just described, the head-foot voltage gradient can exceed the dielectric break-down of air and the flash continues over the surface of the body. A variety of pathophysiological responses result from the current that flows through the body; these will be discussed subsequently.

It is extremely unwise to seek shelter under a tree because objects projecting above the ground attract lightning; the higher the object is, the greater the chance of being struck. In Figure 4.5B the voltage gradient between the tree and subject exceeded the air dielectric breakdown and a side flash struck the subject. Such a side flash is dangerous and can be avoided by seeking shelter elsewhere.

Figure 4.5. In A is shown lightning striking a subject and continuing to ground, passing over the surface of his body due to the head-foot voltage exceeding the dielectric breakdown of air. In B is shown a side flash in which the lightning struck a tree, then the subject thereunder.

TYPES OF CONTACT

Although lightning is an outdoor event, indoor as well as outdoor subjects have been shocked. The indoor shocks result from a lightning discharge to the telephone or power line. In the former case, while using the telephone, an acoustic and electrical shock can be experienced, despite the presence of lightning protective devices. There are cases of telephone shock in which the subject has been thrown physically, probably due to sudden strong acoustic stimulus and/or muscular contractions. In some instances, a telephone shock can produce temporary deafness or unconsciousness; it is rarely, if ever, fatal.

Power-line mediated shocks result from a subject being near or in contact with an appliance connected to the power line when it is struck by lightning. Such shocks can vary in severity, depending on how well the subject is in contact with the appliance and ground. Therefore it is prudent to unplug appliances or keep away from them during a thunderstorm.

Outdoor lightning shocks are the most common and the shock or injury may be due to the step-voltage rise, direct strike or side flash, the former being facilitated by the subject wearing or carrying a metal object (jewelry, cap badge, umbrella, golf clubs, metal fishing rod etc.) or being the most prominent object on the ground.

NATURE OF INJURIES

Although there is no typical lightning injury, Andrews et al. (1992) classified the injury severity as mild, moderate and severe. With mild injury the subject is stunned; but is awake and confused, often with recent memory loss. Frequently there is muscle pain that can persist for several days.

With moderate injury the subject has lost consciousness or is confused, often with paralysis that may last for hours after regaining consciousness. There may be first or second-degree burns and respiratory arrest, which can cause brain damage if cardiopulmonary resuscitation (CPR) is not applied promptly. Tympanic membrane damage, difficulty with fine motor movements and sleep disorders are common.

With severe injury, cardiac and respiratory arrest occur and, if present for many minutes, can result in brain damage. Often there is trauma, bone fractures and intracranial injuries. The prognosis is poor except for those who receive prompt CPR. With severe injury, there is intense vasoconstriction, making it difficult to feel a peripheral pulse, except for the carotid, which is the recommended site for palpation. It is clear that prompt CPR is indicated in moderate and severe injury. No emergency therapy is needed for mild injury.

Jex-Blake (1913) provided the following interesting observations on subjects struck by lightning. He wrote "Very characteristic, or even diagnostic, of lightning-stroke are the so-called 'lightning-figures' often seen on the skin in both fatal and non-fatal cases. These consist of reddish, brown, or purple discolorations of the skin, described in different instances as resembling such things as coral, fronds of ferns, branches of fir-trees, trees, [or] palms."

He also described the burns associated with a lightning strike and wrote:

"Most commonly the burns are of the first or second degree, in the form of streaks that are taken to show the paths followed by the lightning, or in the form of isolated spots or areas. Singeing of the hair growing on these burns is highly characteristic of lightning-strokes, and may take place either with or without any singeing of the clothes in contact with the burnt skin, and also without any burning of the skin itself. Usually the burns are in the shape of bands or ribbons of hardened, discoloured, parchment-like skin, running down the trunk or legs. A very exceptional case is recorded by Cipriano, in which there were burns of the second degree on the neck, chest, and leg, while practically the whole of the rest of the body showed a burn of the first degree; the patient made a slow and difficult recovery. Vincent quotes the case of a woman who was struck and received burns of the first to the third degree from head to foot, and died from exhaustion after six months' suppuration. Less often the burns go deeper, and may even char the underlying bones or the tongue (Clark and Brigham) or the abdominal viscera. In many cases the deeper burns heal readily, if the victim survives; but in others slow healing with much pain and suppuration have been recorded. Patches of hair and skin may be quite torn off. Persons struck or killed by lightning are said to exhale a peculiar or characteristic odour, oftenest described as resembling the smell of burning sulphur or of ozone; other writers compare the odour to such various smells as those of nitrous fumes, gunpowder accidents, acid, dilute sulphuric acid, sulphuretted hydrogen, nitrous acid, ammonia, electric sparks, and lighting matches, and it has also been described as *sui generis*" (of its own kind).

To provide perspective on the type of injury produced by lightning, Andrews and Darvenzia (1989) summarized the most frequently observed symptoms and signs among 221 cases in the literature up to 1987. The most common symptom was loss of consciousness (59%), restlessness, hysteria and dizziness (12%), paresthesias and hypoesthesias (17%), lightning paralysis (41%), upper-limb paralysis (29%), amnesia (46%), cardiopulmonary arrest (20%) with ECG abnormalities in only 10% of the cases. Among the lesser encountered findings were cerebral edema, hyporeflexia, hyperthermia, hypertension, bradycardia, ruptured eardrum and damage to clothing.

Jex-Blake (1913) was one of the first to point out that there are fewer fatalities than the number of lightning strikes. He stated:

"Many more people are struck by lightning than are killed. For example, Jack records an instance in which a church was struck; 300 people were in it, 100 were injured and mostly made unconscious, 30 had to take to their beds, but only 6 were killed. Weber gives an account of 92 people struck in Schleswig-Holstein; 10 were killed, 20 paralyzed, 55 stupefied, and 7 only slightly affected. In 1905 a tent with 250 people in it was struck, and 60 were left on the ground in various states of insensibility; one was killed outright, another breathed for some minutes before dying, the rest recovered. As many as eleven and eighteen persons have been killed by a single stroke of lightning. Vincent mentions a stroke that threw down 1,200 and killed 556 out of a flock of 1,800 sheep."

Finally, Jex-Blake (1913) wrote:

"In a few well recorded instances which are extraordinary almost to the point of being incredible, strokes of lightning have effected amputations. Sycyanko quotes a case occurring in Russia, in which a boy of 12, who had had a flexed and ankylosed right knee for some years, was struck when riding on horseback. He was thrown to the ground and rendered insensible. At the same time the lightning amputated his right leg just below the knee-joint, leaving the patella and the cartilaginous upper end of the tibia intact. The boy made a good recovery; the amputated limb was afterwards found near the spot where he was struck. Vincent refers to a case in which an arm was amputated by lightning, but he did not see it himself. Dunscombe-Honeyball records the fact that he has known lightning to amputate a man's fingers. There seems no reason to doubt that such amputations might be the results of local development of heat by the passage of the lightning through the tissues, with the sudden production of steam or other hot gases in sufficient quantity to blow the limb off."

CONCLUSION

Cox (1992) presented a summary of the precautions that one should take to avoid being injured by lightning. These recommendations are summarized by the author as follows:

1) Seek shelter in a large building or an all-metal vehicle such as a car, but not a convertible. Side flashes may occur from the inside walls of small buildings, especially metal sheds.

2) Avoid tents because the poles may act as lightning rods (attracters of lightning).

3) If indoors, avoid open doors and windows, fireplaces, metal objects and electrical appliances. Do not use the telephone. Disconnect portable telephones and appliances.

4) Keep away from metal objects such as motorcycles, tractors, fences and bicycles. Put down umbrellas or golf clubs and remove headgear if it contains metal, such as a badge. Metal studs in shoes also act as excellent grounding agents.

5) Avoid proximity to power lines, pipelines, fences, ski lifts and other steel structures.

6) Do not stand under isolated trees or on hilltops. Seek shelter in a wood under small trees or saplings (see #7).

7) If totally in the open, avoid being the highest object. If standing, keep feet together to avoid a shock from the potential step (rise) in the ground, but preferably kneel, crouch or roll up in a ball and keep away from haystacks or other isolated objects that project above ground.

8) If in or on the water, go ashore. The head of a swimmer may be the tallest object around in a large expanse of water and may consequently be struck. The mast of a boat may attract lightning and the occupants of may be injured or the lightning may be conducted through the keel and injure swimmers nearby.

9) A group of people should disperse several yards apart so that, in the event of a strike, the least number are injured by ground current or side flashes between persons (see #7).

The author will add a tenth recommendation, namely learn CPR so that you can resuscitate a victim immediately because in cardiopulmonary arrest, time is the enemy.

REFERENCES

Andrews, C.J., Cooper, M.A., Darvenzia, M. et al. Lightning Injuries. Boca Raton FL, 1992. CRC Press, 195 pp.

Andrews, C.J., Darvenzia, M. and Mackerrans, D. Lightning Injury: Chicago 1989 Year Book Publishers.

Cox, R.A.F. Lightning and electrical injury (editorial). J. Royal Soc. Med. 1992, 85:591–593.

Eriksson, A. and Ornehuit, L. Death by lightning. Amer. J. Forensic Med. Pathol. 1988, 9:295–300.

Jex-Blake, A.J. Death by lightning. Brit. Med. J. 1913, March 15:548–552.

Malan, D.J. Physics of Lightning. London 1969. The Universities Press 176 pp.

Ohashi, M. Lightning injury caused by discharges accompanying flashover. Burns 1986, 12:496–507.

Uman, M.A. Lightning. New York, 1969. McGraw Hill 264 pp.

Viemeister, P.E. The Lightning Book. MIT Press 1972.

Chapter 5
RESPONSE TO HIGH-FREQUENCY CURRENT PASSING THROUGH THE BODY

INTRODUCTION

High-frequency alternating current is passed through the body intentionally for electrosurgery and diathermy. In the former case, localized heating is produced at the tip of a hand-held probe to cut and coagulate tissue. In the latter case (diathermy), high-frequency current produces bulk heating to dilate blood vessels and thereby increase the circulation. High frequency current is induced into the body during magnetic resonance imaging (MRI). Radio, television and radar all use high-frequency current which can be encountered accidentally and cause injury. Prior to discussing the various uses and accidents that can result from exposure to a high-frequency source, it is useful to summarize the frequency spectrum; Table 5.1 presents a compilation of designations and services. Not shown in Table 5.1 are special frequencies reserved for armed forces and government use; these are selected frequencies in the range of 510–535 kHz, 25.33–50 MHz, 138–420 MHz, 890–2,900 MHz and 3.1–38.6 GHz. Cellular telephones operate at about 800 MHz.

ELECTROSURGERY

The terminology associated with electrosurgery is not always precise. Terms such as cautery, electrocautery, surgical diathermy and Bovie are in common use. The following definitions of these terms will serve as a guide to use of the correct term. A cautery is a heated rod (like a poker) used to cut and coagulate tissue; its origin is in Arabic medicine. An electrocautery is an electrically heated rod, not unlike a soldering iron, which can indeed be used to cut and coagulate tissue. Surgical diathermy is an older term that is used in the UK to designate the use of high-frequency current in surgery. The term "Bovie" is used almost inter-changeably with an electrosurgical unit because it was W.T. Bovie who built the first practical (spark-gap) electrosurgical unit, that was introduced to medicine in 1928 by Cushing and Bovie (see Geddes et al. 1977 for the history of electrosurgery). Because modern electrosurgical units no longer contain spark gaps or vacuum tubes to generate the cutting and coagulating currents, the correct term for the generator of high-frequency current for surgery is an electrosurgical unit (ESU).

TECHNIQUES

Electrosurgical techniques use 0.5–2 MHz radio-frequency current delivered to the tissue by a hand-held probe to dessicate, coagulate and cut living tissue. Figure 5.1 is a sketch of the three components of an electrosurgical system: 1) the hand-held probe (active electrode), 2) the electrosurgical unit (ESU) and 3) the dispersive electrode. It is at the probe tip, which may or may not be in contact with the tissue, where the three processes (dessication, coagulation and cutting) occur. The ESU is under the control of the surgeon, either by a push-button on the probe or by a double foot switch. The large-area dispersive electrode provides a safe return path for the electrosurgical current. It is important to recognize that heating depends on current density (mA/cm^2) squared and the duration of current flow. Therefore the maximum heating is under the tip of the small-area, hand-held probe electrode. The dispersive electrode is designed to have a large area so that negligible skin heating occurs thereunder.

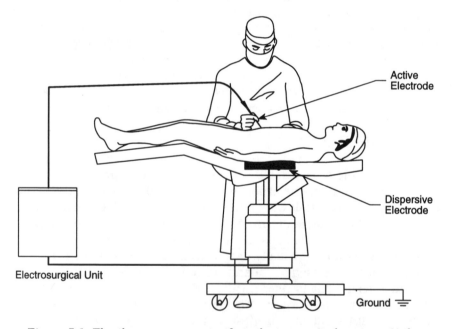

Figure 5.1. The three components of an electrosurgical system: 1) the small-area, hand-held probe (active electrode), 2) the electrosurgical unit (ESU) and 3) the large-area dispersive electrode.

There are hand-held bipolar electrosurgical electrodes that are used for coagulating small vessels. No dispersive electrode is required with such a bipolar electrode. Such bipolar electrodes can be activated by a push button on the electrode handle or by a foot switch.

Table 5.1: Frequency Allocations

FREQUENCY RANGE	DESIGNATION	SERVICE
30–300 kHz	Very Low Frequency (VLF)	Radio location, navigation
30–300 kHz	Low Frequency (LF)	Radio location, navigation and maritime
300–3,000 kHz	Medium Frequency (MF)	535–1605 AM broadcast 1605–3500 Mobile and navigation
3–300 MHz	High frequency (HF)	1.8–2 Aviation 3.5–4 Amateur 4–5.95 Mobile, aeronautical 5.95–6.25 SW broadcast 6.20–7.0 Aeronautical, marine 7.0–7.3 Amateur 7.3–9.5 Aeronautical, marine 9.5–9.75 SW broadcast 9.75–11.7 Aeronautical, marine 11.7–11.975 SW broadcast 11.975–14.00 Aeronautical, marine 13.56 Diathermy 14.0–14.35 Amateur 14.35–15.1 Aeronautical 15.1–15.45 SW broadcast 15.45–17.7 Space research 17.7–17.9 SW broadcast 26.96–27.23 Citizens band 27.12 Diathermy
30–300 MHz	Very high frequency (VHF)	54–88 TV channels 2–6 88–108 FM broadcast 119.75–132 Aeronautical (Tower) 174–216 TV channels 7–13 225–390 Radar (P) 390–1550 Radar (L)
300–3000 MHz	Ultrahigh frequency (UHF)	462.525–467.475 Citizens band 470–890 TV channels 14–83 915.0 Diathermy (not in U.S.) 1550–5200 Radar (S) 2450 Diathermy
3–30 GHz (1 GHz = 1000 MHz)	Super high frequency (SHF)	5.2–10.9 radar (X) 5.80 Diathermy 10.9–36 radar (K) 22.125 Diathermy
30–300 GHz	Extremely high frequency (EHF)	36–40 radar (Q) 46–56 radar (V)
300–3000 GHz	--------	56–100 radar (W)

TYPES OF CURRENT AND TISSUE RESPONSES

Two types of radio-frequency current (unmodulated and modulated) are used to induce the three types of tissue response (dessication, coagulation and cutting). Moreover, there are two techniques for applying the hand-held probe, also called the active electrode or pencil: one brings the tip of the probe in contact with the tissue, then the ESU is activated. The other technique brings the tip of the probe in close proximity to the tissue; then the ESU is activated and an arc carries the current to the tissue. Which technique is used depends on the desired tissue response.

Figure 5.2 illustrates the current waveforms for cutting and coagulation. In general, cutting is achieved with continuous, i.e. unmodulated current; Figures 5.2A, B illustrate the type of current used for cutting. The waveform in Figure 5.2A is produced by the older vacuum-tube (Bovie) units. The radio frequency cutting current is delivered in 120 second, half-sinusoidal bursts. In the newer, solid-state units, the current is more constant as shown in Figure 5.2B.

The coagulating current waveforms are shown in Figures 5.2C, D. Note that in both cases the duty cycle is short. Figure 5.2C illustrates the coagulating waveform provided by the older spark-gap (Bovie) generators and the bursts of radio frequency current are delivered at 120/sec. Figure 5.2D illustrates the coagulating waveform provided by the newer solid-state units which deliver the bursts at typically 20,000/sec.

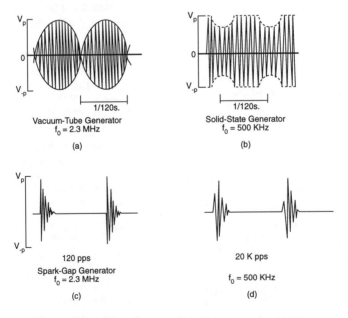

Figure 5.2. Waveforms of cutting current (A,B)
and coagulating current (C,D).(Courtesy of J.A. Pearce)

Dessication

Dessication (Figure 5.3A), i.e. drying, is produced by placing the tip of the probe in contact with the tissue to be desiccated and activating the ESU. Low current (of either type) can be used; no arc is formed and the fluid is driven from the tissue surrounding the probe.

Coagulation

Coagulation employs modulated current (Fig 5.2C, D) and several different techniques are employed with the active electrode. With spray coagulation (Figure 5.3B), sometimes called fulguration (fulgur = lightning), the active electrode is brought close to the tissue and the ESU is activated. Arcs travel from the tip of the probe to the tissue, each arc striking a different site, the result being a more-or-less circular area of coagulation. This technique is used to remove skin blemishes and sometimes to close very small bleeding vessels. Alternately, the active electrode is placed in contact with the bleeder (Figure 5.3C) and the ESU is activated; virtually no arcing occurs.

Another technique for closing larger blood vessels was developed by Ward (1925) and is shown in Figure 5.3D. The bleeder is grasped with a hemostat or forceps to arrest bleeding and the tip of the active electrode is touched to the hemostat; then the ESU is activated. No arc is formed and either type of current can be used to seal the bleeder. The surgeon's gloves provide insulation for the hands. There are special forceps in which the two jaws constitute a bipolar electrode. The term used in this instance is bipolar coagulation and the current flow is localized to the two electrodes.

Surgical cutting (Figure 5.3E) employs unmodulated radio frequency current of the type shown in Figure 5.2A, B. The tip of the probe is brought to the tissue surface and the ESU is activated. The tissue vaporizes, producing a scalpel-like cut; the permissible speed of cutting depends on the current intensity. However the lips of the incision usually bleed and coagulation current is used to arrest the bleeding. In many ESUs, it is possible to combine (blend) cutting and coagulating current electrically so that coagulation accompanies cutting.

In laparoscopy, the surgical probe is passed into the endoscope, as shown in Figure 5.3F for operation within the urinary bladder or urethra. This type of electrosurgery uses higher current and, due to the capacitance between the conductor in the probe, its insulating sleeve and the surrounding tissue there is some heating of tissues along the length of the probe. This effect can be demonstrated if a dispersive electrode is placed on the arm and the active electrode held in the hand, the tip not touching anything. With a high ESU output, the hand feels a sensation of warmth; this is really capacitive diathermy heating.

In Figure 5.3G is shown an active electrode with two push buttons; one allows delivery of cutting current and the other permits delivery of the coagulating current. Such hand-held probes come packaged, sterilized and are disposable items. With most ESU units, a dual foot switch controls delivery of these currents to the hand-held active electrode.

Current Levels

Table 5.2 presents a summary of current levels (rms amps) used in various electrosurgical procedures, along with typical durations of activation of the ESU. It is important to recognize that the ESU is activated many times during a surgical procedure.

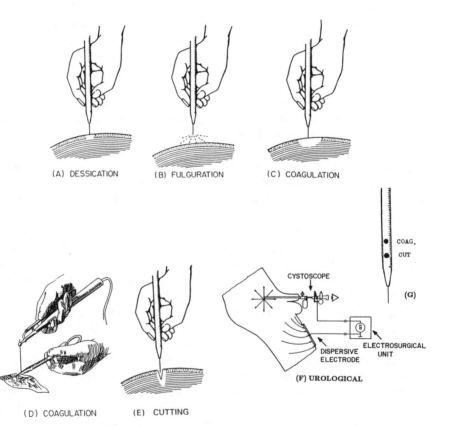

(A) DESSICATION (B) FULGURATION (C) COAGULATION

(D) COAGULATION (E) CUTTING (F) UROLOGICAL

CYSTOSCOPE

DISPERSIVE ELECTRODE ELECTROSURGICAL UNIT

COAG. CUT (G)

Figure 5.3.   Dessication (A), fulguration or spray coagulation (B), coagulation with the probe in contact with the tissue (C), coagulation facilitated with a hemostat (D), cutting (E) and urologic technique (F). In G is shown an active electrode with two pushbuttons; one is for cutting current and the other is for coagulating current.

Table 5.2: Current used in Electrosurgical Procedures[*]

Proce-dure	Current[a] (mA)			Duration of activation[a] (sec)		
	Min.	Max.	Avg.	Min.	Max.	Avg.
Transurethral resection						
Cut	239 (162)	407 (297)	297 (200)	1.6 (.68)	3.8 (2.3)	2.1 (.69)
Coag.	179 (78)	419 (400)	256 (88)	1.4 (.5)	5 (7.6)	1.9 (.7)
Laparoscopic tubal ligation						
Cut	126 (120)	430 (290)	239 (135)	1.7 (.58)	5.4 (4.9)	2.6 (3.2)
Coag.	61 (57)	118 (80)	86 (70)	3.2 (.31)	26 (20)	10 (7.4)
General surgery[b]						
Cut	238 (188)	340 (101)	281 (147)	2 (2)	7.6 (11)	2.2 (1.8)
Coag.	146 (94)	267 (157)	198 (114)	4.7 (5.2)	11 (7.8)	6.5 (5.2)

[a] Numbers in parentheses are standard deviations.
[b] General surgical procedures include prostatectomy, laparotomy, thoracotomy, hip pinning, hysterectomy, nephrectomy, and D&C.
[*] Courtesy of J. De Rosa, NDM Co., Dayton, OH.

DISPERSIVE ELECTRODE

The dispersive (sometimes called ground or indifferent) electrode provides a safe return path for the electrosurgical current. The electrode is large enough so that there is negligible heating of the skin under it. Typically, the dispersive electrode is placed over a fleshy part of the body; it is never placed over a bony prominence because there is the risk of uneven current distribution which could produce a hot spot.

There are two types of dispersive electrode: 1) conductive and 2) capacitive. The conductive type establishes an ohmic contact with the subject; the capacitive type does not. The capacitive electrode consists of a metal plate covered by insulating film (dielectric). In this way one "plate" of the capacitor is the subject, the other is the metal within the electrode.

CONDUCTIVE DISPERSIVE ELECTRODES

The first dispersive electrodes consisted of a large-area, bare, dry metal plate on which the patient lay, shown schematically in Figure 5.4A. Variable contact area led to the use of flexible metal foil, insulated on the back and surrounded by an adhesive perimeter which allowed placement of the electrode at any convenient body site. The temperature distribution under a dry metal-plate electrode on human skin was reported by Pearce et al. (1978) and Geddes et al. (1980). Figure 5.5A illustrates the skin temperature distribution just after the removal of a dry metal-foil dispersive electrode.

To improve contact with the skin, an electrolytic gel is used in some metal-foil dispersive electrodes; Figure 5.4B shows the principal components. A peel-off cover protects the electrode during storage. The skin temperature distribution of a typical gelled metal-foil electrode was reported by Pearce et al. (1978) and Geddes et al. (1979); Figure 5.5B illustrates the skin-temperature distribution after removal of such an electrode.

In some dispersive electrodes, the electrolytic gel is incorporated into a foam pad that bridges the gap between the metal foil and skin; Figure 5.5C illustrates the essential components. A peel-off cover protects the electrode in storage. The temperature distribution, is similar to that for the gelled metal-foil electrode (Fig 5.4C).

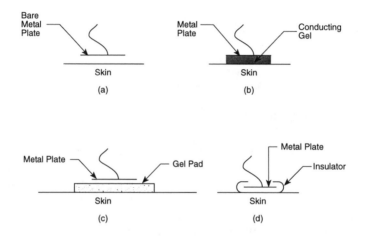

Figure 5.4. Dispersive electrode types.

Advances in adhesive technology resulted in the ability to include an electrolyte and thereby create a conducting adhesive dispersive electrode. Such an electrode consists of a metal foil, backed by an insulator with an adhesive perimeter. The conducting adhesive sticks to the metal foil and a peel-off label exposes the conducting adhesive which embraces the skin well. Such electrodes are in widespread use.

Irrespective of design, the conductive dispersive electrode makes ohmic contact with the subject. In addition, the current distribution under such an electrode is not uniform, the current density under the perimeter being several times higher than under the center of the electrode. Because heating depends on current-density squared, the skin under the perimeter of a conducting electrode is warmer than the skin under the center of the electrode.

Pearce et al. (1979) called attention to the fact that the electrode-perimeter, skin-heating could be reduced by increasing the perimeter-to-area ratio. They compared three electrode shapes (circular, square and figure eight) and found the skin temperature rise decreased with an increase in the perimeter-to-area ratio. This fact has not been exploited in dispersive electrode design; but there is one electrode that was notched to improve conformability; therefore its perimeter-to-area ratio is higher than if it were not notched.

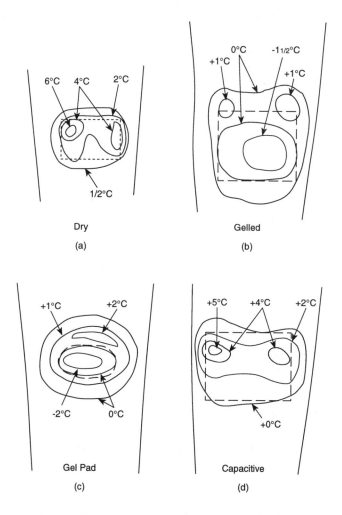

Figure 5.5. Skin temperature distribution under a dry, metal-foil electrode (A), a gelled metal-foil electrode (B), a gel-pad electrode (C) and a capacitive electrode (D).

Handbook of Electrical Hazards and Accidents

CAPACITIVE DISPERSIVE ELECTRODE

Because the frequency (f) of electrosurgical current is high, it is possible to cover the metal foil with a thin insulating (dielectric) film; Figure 5.4D illustrates the principle. The capacitance (C) depends on the electrode area (A), dielectric constant (k) and inversely with the thickness (t) of the dielectric film, $C = kA/t$. The size (area) of the electrode is chosen so that the reactance ($1/2\pi fC$) of the capacitance is sufficiently small that an adequate return path is established with the subject. As with the other dispersive electrodes, an insulating back and adhesive perimeter are provided and a peel-off cover protects the dielectric surface during storage.

Two features distinguish the capacitive electrode from all other types: 1) there is no ohmic contact with the subject, and 2) the skin-temperature distribution is more uniform than that for conductive dispersive electrodes, this being the nature of a capacitive interface. The skin-temperature distribution for capacitive dispersive electrodes was reported by Pearce et al. (1980); Figure 5.5D presents a typical example.

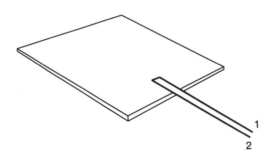

Figure 5.6. Patient-sentry system, consisting of two conductors (1, 2) connected to the dispersive electrode. Within the ESU is a circuit that prevents delivery of electrosurgical current unless there is continuity measured between conductors 1 and 2.

PATIENT SENTRY

To minimize accidents resulting from an electrode applied to the patient but not plugged into the ESU, the patient sentry system has been invented. The dispersive electrode (Figure 5.6) is connected to the ESU by two conductors (1, 2). A circuit within the ESU prevents delivery of electrosurgical current unless there is continuity between conductors 1 and 2. In the earlier ESUs, a relay was held closed by the continuity measured between conductors 1 and 2. The patient sentry was first incorporated into the Bovie CSV units. Newer ESUs use an SCR (silicon controlled rectifier) to allow delivery of electrosurgical current only when continuity is verified.

Although a useful safety method, such dispersive electrodes with two conductors can only be connected to an ESU with the patient sentry feature. Such a patient sentry provides no information on whether or not the dispersive electrode is applied to the subject.

PATIENT RETURN MONITOR

A different type of electrode monitor was developed by Valleylab Inc. and uses a split dispersive electrode, as shown schematically in Figure 5.7. When the electrode is applied to the patient, a circuit in the ESU monitors the impedance between conductors 1 and 2. When the split electrode is on the patient, the impedance between electrodes 1 and 2 is low. If the ESU senses a high impedance between conductors 1 and 2, it means that the dispersive electrode has not been applied properly, or has not been applied to the patient; therefore the ESU is inhibited from delivering current. Note that this type of monitor identifies 1) connection of the electrode to the ESU, 2) proper application of the electrode to the patient and 3) continuity of the two conductors leading to the ESU.

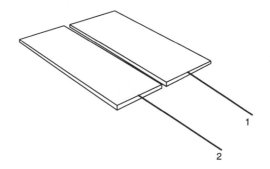

Figure 5.7. Split dispersive electrode in which the proper application to the patient is identified by a low impedance appearing between conductors 1 and 2.

PERFORMANCE STANDARDS

As stated previously, the function of a dispersive electrode is to provide a safe return path for the electrosurgical current. The area of the electrode must be large enough so that excessive skin heating does not occur. There is a performance standard (AAMI HF18-R-2/93) being developed. Briefly it states that the skin-temperature rise must not exceed 6°C for a 700 mA (rms) current applied for 60 sec. The standard should be consulted for the methods for testing.

SKIN BURNS

As stated previously, the heating at a skin-electrode site depends on the current density squared and the duration of current flow. Pearce et al. (1983) introduced this concept, calling it the energy-density factor. Before discussing the burn severity in relation to the energy-density factor, it is useful to identify events that occur in the absence of a satisfactory dispersive-electrode contact with the patient. In this case, the return path for the electrosurgical current will be any grounded object that the patient contacts. An example is shown in Figure 5.8, showing that the return path is via the ECG monitoring electrodes causing current (I_m) to flow, and via any other contact between the patient and the grounded operating table, causing current (I_R) to flow. Because these areas of contact are not large, the current density will be high and burns can result at these sites. Such alternate-path currents are by no means uncommon in accident cases.

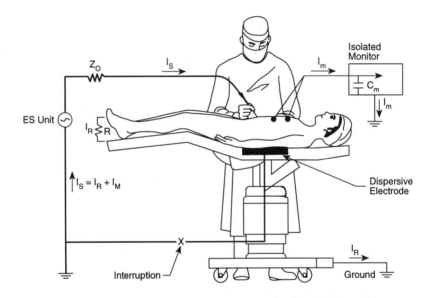

Figure 5.8. Alternate current pathways for electrosurgical current with a grounded-output electrosurgical unit and an interrupted (X) dispersive electrode cable. The surgical current (I_s) returns via any contact between the patient and operating table (R) and via the capacitance to ground (C_m) in a typical isolated monitor.

Another situation that produces a burn results when the dispersive electrode is poorly applied and does not make complete contact with the patient. Buckling (tenting) of the electrode or partial contact resulting from reposturing the patient can result in a reduced area of contact and a burn.

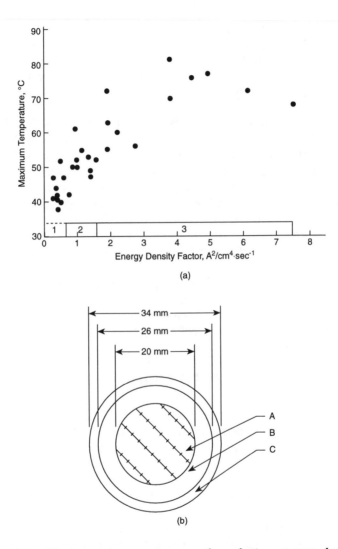

(a)

(b)

Figure 5.9. Maximum temperature vs the relative energy density factor (A). The numbers on the abscissa identify 1st, second and 3rd-degree burn regions. In B is shown the lesion produced by a high current (0.5 A) applied to a 2-cm electrode for 60 sec. A = brown depressed necrotic area (third-degree burn); B = yellowish-white area (third-degree burn); and C = red hemorrhagic zone (second-degree burn). (Redrawn from Pearce et al. 1983).

EXPERIMENTAL BURN STUDIES

Pearce et al. (1983) carried out controlled burn studies on pigs using 500 kHz electrosurgical current delivered to circular disk electrodes 1, 2, 4 and 8 cm applied to the backs of pigs. Pig skin is a good analog of human skin. The skin temperature and burn severity were determined for different currents (300–700 mA) and times (30–60 sec). The skin temperature, determined thermographically, was plotted versus the energy density factor (seconds multiplied by the square of the current density in amps/cm^2 of electrode area). Figure 5.9A illustrates the result; the numbers (1, 2, 3) along the abscissa identify the burn degree. In general, with an energy-density factor in the range of 0.7 to 1.6 the maximum skin temperature below the electrodes was between 49 and 55°C, with single or multiple rings of second-degree burns located just inside or beyond the rim of the electrode. At all sites exposed to higher energy density (1.60–7.50), the maximum skin temperature beneath the electrodes was 55–81°C, and severe burns were produced with white to brownish, dry, firm, third-degree burns surrounded by peripheral rings of second-degree burns. No significant skin damage was produced with a skin temperature less than 45°C, representing an energy density factor of 0.75.

Figure 5.9B is a sketch showing the distribution of skin damage under a small circular metal plate (2 cm) that carried 0.5 amp of electrosurgical current continuously for 60 sec. Note that there are both second and third-degree burns, reflecting the nonuniform current-density distribution under the electrode.

ALTERNATE CURRENT PATH BURNS

A serious hazard to a patient occurs when the return path to the ESU is interrupted. This can occur if the dispersive electrode is not applied or if applied, but not connected to the ESU, or if properly applied to the patient and connected to the ESU, but there is an interruption in the conductor from the dispersive electrode, as shown by X in Figure 5.8. Because one side of the ESU is essentially at ground potential for the radio frequency current, the return path, in the absence of a dispersive electrode, is any part of the body that contacts a grounded object; this will be the site of a burn. This situation is common in many electrosurgical burns which are undetected because the patient is draped and anesthetized.

NON-ELECTROSURGICAL SKIN LESIONS

Not all injuries to the skin of patients undergoing long-duration surgical procedures are attributable to electrosurgery. For example, the pressure on the body being supported by the operating table can be high enough over bony prominences to exceed the tissue capillary pressure, which is about 30 mmHg. High local tissue pressure reduces capillary blood flow and if present for long enough, tissue injury occurs as demonstrated by the following examples.

Lawson et al. (1976) reported the incidence of postoperative occipital alopecia (loss of hair on the back of the head). He stated, "Postoperative alopecia is a minor complication of surgery but a cosmetic disaster to the patient. Over a 3-year period, 60 cases of occipital alopecia were discovered in patients following open-heart surgery and 5 cases on other surgical services. In contrast to previous reports, 20 patients had alopecia one year later, presumed to be permanent. Extensive operations, with prolonged recovery and elective overnight mechanical ventilation, were common to all. Retrospective analysis and prospective studies clearly demonstrated that localized scalp pressure was the cause of the alopecia and that the duration of pressure determined the extent of the damage. Moving the patient's head at regular intervals during operation and recovery eliminated the alopecia. The type of head rest used did not modify the development of alopecia. Electrical injury and the use of heparin, hypothermia, electrocautery, or hypotension were eliminated as possible causes. Conclusive evidence correlating perioperative events with the formation of pressure sores in man has not been previously reported."

The pressure distribution under supine subjects is of concern when the same position is maintained for a long time. If there are sites where the pressure exceeds about 30 mmHg, the blood flow through capillaries at such sites is impaired and if this situation lasts long enough, local tissue injury and death (necrosis) occurs, resulting in what is known as bedsores or pressure sores. Various types of pressure mapping devices have been created (see Babbs et al. 1990) to identify the high-pressure sites in subjects lying on sleep surfaces; Figure 5.10 illustrates one such map of equal-pressure contours in a supine subject; the smaller the enclosed contours, the more localized the pressure. Note the high-pressure regions under the head, shoulder blades, buttocks, thighs, calves and heels; these are the sites most prone to pressure-induced injury; dispersive electrodes should never be placed on these sites because of the poor blood perfusion, impairing the ability to carry away heat.

The foregoing should allow an investigator to decide if postoperative tissue injury is due to pressure necrosis or a burn from electrosurgical current. Lesions due to elevated skin temperature under a dispersive electrode are obvious on removal of the electrode; however, with the passage of time (e.g. a few days), the full extent of the injury is manifest.

Lesions due to pressure-induced capillary damage may take time to manifest themselves.

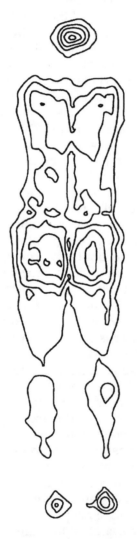

Figure 5.10. Isopressure contour map of a supine adult subject lying on a sleep surface. Higher pressures are under the head, shoulders, sacrum, thighs, calves and heels—the regions that support the body.

TYPICAL ACCIDENTS

A report by Becker et al. (1973), at a time when only conductive dispersive electrodes were used, provides information on the types of electrosurgical accidents that can occur. The report states: "During electrosurgery, nine patients were burned at the sites of electrocardioscope electrodes. Causes were: 1) broken patient-plate ground wires detectable only by conductivity measurements; 2) defective silicon-controlled rectifiers in the sentry modules; 3) a design fault apparent only when the foot switch of one company was used with the electrosurgery machine of another; 4) improper use of the active (knife) electrode; 5) capacitive coupling of radio frequency current in electrocardioscope cables; and 6) radio frequency current division (alternate path)." From this report it is clear that there was failure to provide a return path for the electrosurgical current, equipment failure and the use of incompatible components.

A case report by Irnich (1986) is difficult to explain; he stated:

"Electrosurgery was used during tonsillectomy of a young girl with no effect, motivating the physician to increase the intensity. Irregular current flow led to the change of the neutral electrode (a stiff brass plate which had been covered with gel) from the left upper arm to the right arm. It was discovered that an oval (1×2 cm) skin lesion had been created. Because problems still persisted, it was tried for a third time using the left thigh. Failing once again, it was concluded that the electrosurgery apparatus was defective.

"When the apparatus was judged defective, a new electrosurgical unit was brought. With the introduction of this new device, no further problems were encountered - proof that the old machine really was defective."

Increasing the ESU output above a typical value when inadequate cutting or coagulation occurs is a danger sign that often goes unheeded. The intermittent current and the skin burn may have indicated poor contact between the brass plate (neutral or dispersive) electrode. The size of the lesion 1×2 cm certainly indicates a small area of contact and a high current density. Finally why the problem was solved by the new ESU is not explained because the presence of a skin burn indicates that the first ESU was functioning. It is useful to note that during surgery, the patient is unconscious and draped; a severe burn can occur without knowledge of the patient or surgeon and is usually discovered in the recovery room. The drapes make it difficult to perform a visual or manual check of the dispersive electrode and its adherence to the skin.

MUSCLE CONTRACTION

It is by no means rare that a surgeon encounters muscle contractions while performing electrosurgery near a motor nerve or muscle. Moreover,

ventricular fibrillation has also occurred during electrosurgical procedures. Based on knowledge of stimulation (Chapter 2), it would be predicted that radio frequency current would be unlikely to stimulate such tissues because of its high frequency. This paradox requires an explanation.

Foster and Geddes (1986) devised a simple experiment designed to
test the ability of electrosurgical current to stimulate the dog sciatic nerve
with and without an arc at the tip of the electrosurgical probe, which was
remote from the nerve, as shown in Figure 5.11. The active electrosurgical electrode was placed over a saline-soaked sponge in a metal dish
which was connected to a saline-soaked, gauze-covered electrode applied
to the exposed sciatic nerve of an anesthetized dog. A conventional
dispersive electrosurgical electrode provided the return path to the ESU.
Thus, the electrosurgical current could be applied with or without an arc
by selecting the position of the tip of the active electrode (e.g., above the
sponge or advanced into it), as shown in Figure 5.11.

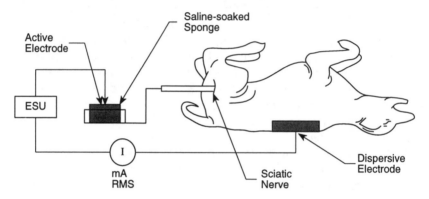

*Figure 5.11. Arrangement of equipment to apply electrosurgical
current to a nerve with and without an arc at the active electrode
applied to a saline-soaked sponge in a metal dish. (Redrawn from
Foster & Geddes, 1986).*

Stimulation of the sciatic nerve causes contraction of the gastrocnemius muscle and rotates the foot (downward) around the ankle joint.
Therefore, if any pulsating current flowed in the circuit, the sciatic nerve
would be stimulated, thereby contracting the gastrocnemius muscle.

Two types of experiment were performed. One consisted of applying
electrosurgical current with no arc, i.e., the tip of the active electrode was
plunged into the sponge before the ESU was activated; the other employed activation of the ESU with the tip of the active electrode just above
the sponge and slowly advancing it until an arc was truck. Thus, the current that flowed represented the no-arc and arc conditions. The rms current was measured in both cases with a thermocouple-type ammeter.
Table 5.3 summarizes the results and shows that with no arc, in only one

case was there a trace of muscle contraction. In all cases with an arc, muscle contraction occurred.

Table 5.3: The Stimulating Capabilities of Cutting and Coagulating Current

Instru-ment	Mode	Current [mA(rms)]	Contraction	
			Arc	No Arc
1	Coag.	200	Yes	No
	Cut	500	Yes	No
2	Pure cut	200	Yes	No
	Spark gap coag. (moderate hemostasis)	600 200	-- Yes	No
	Spark gap coag. (marked hemostasis)	500 100	-- Yes	No
	Spark gap coag. (max. hemostasis)	250 <100	-- Yes	Trace

Source: Foster and Geddes, *Med. Instrum.* 20(b): 335–336 (1986).

It has been shown by Pearce et al. (1986) and Tucker et al. (1984) that the presence of an arc in the path of electrosurgical current alters the frequency spectrum markedly, producing low-frequency components capable of stimulation. That the cause of muscle contraction during electrosurgery is due mainly to stimulation of motor nerves comes from two sources: 1) the muscle contraction can be distant from the site of the active electrosurgical electrode and 2) muscle relaxants markedly reduce the muscle contractions, as reported by Prentiss et al. (1965).

Geddes and Moore (1990) reported the following accident that resulted from muscle contractions during electrosurgery:

"A 32-year old white male with tender enlarged lymph nodes in the left posterior cervical triangle presented to the office of a general surgeon. There was also a 25 to 30-pound weight loss over a two to three-month period. He was hospitalized for diagnostic studies which were negative. Suspecting possible Hodgkins disease, he was taken to the operating room for removal of the largest node. When the surgeon encountered a small bleeder in the subcutaneous fat, he used a standard, solid-state electrosurgical unit in the monopolar mode with a coagulation at a setting between 3 or 4 to stop the bleeding. As the bleeder was approached with the active electrode to perform coagulation, the patient experienced a severe jolt which he likened to being shocked by an electric outlet. The operative report made no reference to the incident or the electrosurgical unit, stating only that "A nerve, presumably of the cervical plexus was anesthetized and retracted posteriorly and the subjacent node was dissected out. The discharge summary, however, noted that during this operation, while in the subcutaneous tissue under local anesthesia, a standard ESU was used to coagulate a small bleeding blood vessel and this caused a rather violent jerk in the patient's body that he said he felt all the way down into his left leg and left arm. The ESU was used no further during the operation. As the operation continued, the accessory nerve was noted to be near the area and it was presumed that the stimulation of a nerve is what caused this violent spasm."

The physician's office records noted that "electric coagulation of a bleeder resulted in a fairly violent spasm of what I believe is probably trapezius and sternomastoid muscles due to stimulation of the underlying spinal accessory nerve."

The surgeon was concerned that the ESU had malfunctioned. Although the machine used in the procedure was not identified, all machines were checked approximately two months after the operation and all were reported to be functioning normally. The disposable active electrode and grounding pad were not retained.

Subsequent to the operation, the patient complained of pain in the neck, left arm and shoulder. When seen in the physician's office three days later, he was unable to completely abduct his left arm. An electromyographic study 18 days after surgery revealed "moderately severe denervation of the rhomboids, levator scapulae, supra-spinatus, infra-spinatus, and trapezius muscles," prompting the neurologist to note that the results "may provide evidence of involvement of the nerve to the trapezius, dorsal scapular nerve, and supra-scapular nerve. This may be secondary to partial injury to the upper brachial plexus on the left.

"Approximately one month after surgery, the surgeon was concerned that the patient exhibited diffuse atrophy of his deltoids and other shoulder girdle muscles. Subsequent orthopedic treatment resulted in a conclusion of a dual lesion: brachial plexus injury and cervical disc degeneration at C5–6, the latter resulting in anterior intervertebral disc excisions and fusion from C4 through 7. The orthopedic surgeon felt that the patient "suffered an electrical injury to the brachial plexus from direct spread of current. In addition, he doubtless had degenerative disease in his cervical spine which was not symptomatic until the violent jerking of his neck occurred. After that, the symptoms persisted until the surgery was performed on the cervical disc."

From the foregoing, it appears that the active electrosurgical electrode was near a motor nerve or plexus and that the presence of the arc provided the rectification that demodulated the radio frequency current which produced low-frequency components that resulted in stimulation. Of the two types of electrosurgical current, the coagulating current, which is delivered in bursts to achieve effective coagulation, is the most likely to stimulate if an arc is struck. However cutting (unmodulated) radio frequency current has the potential for stimulation if an arc is struck.

VENTRICULAR FIBRILLATION

The foregoing has shown that electrosurgical current can stimulate excitable tissue such as motor nerves; therefore it follows that it can stimulate cardiac muscle. As described in Chapters 2 and 3, repetitive stimulation of the ventricles can produce ventricular fibrillation. The first such an event in man due to electrosurgical current was reported by Hungerbuhler et al. (1974); the report reads:

"A 24-year-old woman was scheduled for elective surgical correction of an asymptomatic secundum-type atrial septal defect. Preoperatively, a central venous catheter and two intravenous catheters were inserted percutaneously. A left radial arterial catheter, connected to an electronic pressure transducer (Hewlett-Packard) was also inserted. An electrocardiographic oscilloscope, with an electrocardiographic pad placed under the back, was used. An additional electrocardiogram was monitored with extremity leads connected to a system designed for amplification of low-level electronic signals originating within the body. A patient grounding plate was placed under the right buttock; the ground clip was attached and then plugged into the electrosurgical unit.

"During the initial part of the surgery, the electrocautery was used to cauterize bleeding vessels in the subcutaneous fat, which resulted in the usual amount of high-frequency interference on the oscilloscope. The auscultatory esophageal rhythm remained regular.

"However, 5 to 10 seconds after application of the electrosurgical knife in the cutting mode to the sternal periosteum, the auscultatory rhythm was noted to be irregular and then absent. Simultaneously, an increased electrical high-frequency interference appeared on the electrocardiographic oscilloscope. When the high-frequency interference cleared, the patient was noted to be in ventricular fibrillation, and had a blood pressure of less than 50 mm Hg systolic.

"The patient was immediately resuscitated over the next 70 seconds with external cardiac massage, lidocaine (150 mg total intravenous dose), phenylephrine (Neosynephrine) hydrochloride (2 mg given intravenously), and direct-current electroshock (200 watt-seconds) with paddles applied to the thorax.

"Arterial blood gases at this time measured pH, 7.53; PCO_2, 26 mm Hg; PO_2, 255 mm Hg, with FIO_2, 1.0 mm Hg[?]. A normal sinus rhythm returned promptly and the blood pressure increased to 100/60 mm Hg.

"Since the cause of ventricular fibrillation was not apparent and the recovery was rapid with no further cardiac instability, it was elected to continue with the surgical procedure. The electrosurgical knife was again applied to the sternal periosteum in the cutting mode, which resulted in a second episode of ventricular fibrillation. The patient was immediately resuscitated with prompt return to sinus rhythm and normal blood pressure. At this time, the ground clip to the ground plate was disconnected and examined. A paper remnant torn from a previously used disposable ground plate was discovered wedged into the clip, but contact appeared adequate. Further, the male ground plug to the unit itself was noted to be loose. The unit was replaced with another electrosurgical unit (the same model) with the same ground plate and a new clip."

Becker et al. (1973) reported accidental ventricular fibrillation in dogs during electrosurgery. How this event could occur was investigated by Geddes et al. (1975, 79) using dogs with a saline-filled catheter in the right ventricle connected to a strain-gauge pressure transducer. A dispersive electrode was placed under the dog in dorsal recumbency. The active electrode was used to cut the skin prior to a thoracotomy and no cardiac arrhythmia occurred. The dispersive electrode was then disconnected and immediately ventricular fibrillation occurred when ESU current was delivered to the active electrode applied to the chest.

Figure 5.12 is a record of the ECG and femoral artery pressure of a dog before and during the application of (vacuum tube) cutting current (2.3 M Hz) to the active electrode on the chest, the dispersive electrode being disconnected. Note the immediate occurrence of ventricular fibrillation.

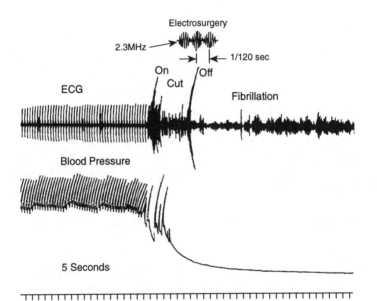

Figure 5.12. Ventricular fibrillation in the dog caused by electrosurgical current. (By permission of J. Bridges).

The foregoing human and animal studies demonstrate clearly that ventricular fibrillation can occur during electrosurgery if current passes through the heart. It has been well documented that electrosurgical current contains low-frequency components that can produce ventricular fibrillation. The most hazardous situation exists when the dispersive electrode is disconnected and the return path for the current includes a grounded, fluid-filled catheter or catheter electrode in the heart or vascular system.

TRACHEAL TUBE FIRE

A case of ignition of the endotracheal tube during elective surgery was reported by Simpson and Wolf (1986). They wrote:

"A 4-yr-old boy was admitted for elective adenoidectomy and tonsillectomy. His past medical history was negative except for recurrent tonsillitis, the last episode having occurred 3 to 4 weeks prior to surgery. Past surgical history included repair of an inguinal hernia 2 months before, under general anesthesia, without complications. The patient was taking no medications and had no allergies. He weighed 17 kg and his physical examination was normal except for large, opposed tonsils touching at the midline. Preoperative blood chemistry, complete blood count, prothrombin time, plasma thromboplastin time, and urinalysis were normal. Atropine, 0.3 mg im, was given 1 h prior to surgery.

"Intraoperative monitors included an oxygen analyzer, left chest precordial stethoscope, blood pressure cuff, electrocardiogram, and rectal temperature probe. Anesthesia was provided using a Bain Circuit. A warming blanket was used to maintain normal temperature.

"Anesthesia was induced by inhalation of O_2, N_2O, and halothane and was uneventful. An i.v. infusion was then started. The trachea was automatically intubated with a 4.5 mm ID polyvinyl chloride (PVC) endotracheal tube. Lubricant was not used on the endotracheal tube. Breath sounds were equal bilaterally in all four quadrants. The endotracheal tube was fixed with the aid of a Davis-Crow mouth gag. Endotracheal tube position was again confirmed by equal breath sounds bilaterally. The eyes were protected with tape. The patient breathed N_2O 3 l/min, O_2 3 l/min, and 1% halothane, by controlled ventilation. A moderate retrograde leak of gases was noted around the tube at the larynx.

"The surgeons performed the adenoidectomy and right tonsillectomy without incident then began to use a suction electrocautery to control the bleeding in the right tonsillar fossa. The electrocautery was set at 35-watts coagulation in the "spray" mode. After approximately 30 sec of cautery, a fire erupted in the pharynx that "blow-torched" toward the lips. Breath sounds were immediately lost and increased airway pressure was noted. The fire was extinguished with saline, the pharynx was suctioned, and the endotracheal tube was immediately removed. The tube was noted to be melted and charred externally for 2 cm, midway between the distal tip and the adaptor, and fused for 1 cm at that point with 100% occlusion. From the point of fusion distally, the tube was blackened internally. The trachea was immediately reintubated with a 5.0 mm ID PVC, uncuffed, endotracheal tube. With a fractional inspired O_2 concentration (F_iO_2) of 0.99 and 1% halothane, pH, was 7.43, Pa_{CO2}, 28 mmHg, and Pa_{O2}, 575 mmHg. On direct examination the mucosa of the posterior tongue, uvula, and hypopharynx were noted to be erythematous and charred. Rigid bronchoscopy using a 4.0 mm×30 cm bronchoscope was performed. The cords were not burned, but were edematous. The mucosa of the anterior trachea, carina, and left mainstem bronchus was erythematous and charred in some areas. This was probably caused by a "blow-torching" of the fire downward. The burns were mostly anterior and not circumferential. After bronchoscopy, the trachea was reintubated orally with a 5.0 mm ID PVC, uncuffed, endotracheal tube under direct vision. Breath sounds were again equal and clear bilaterally. The patient was given dexamethasone 2 mg iv.

"While endotracheal tube fires ignited by lasers have been reported, the only previous case of an electrosurgically ignited PVC endotracheal tube fire was reported by Rita and Seleny, (Anesthesiology 1982; 56:60-61) who reported ignition of an endotracheal tube during the use of a urethral resectoscope for laryngeal surgery. Although the resectoscope uses electrosurgical cutting, it is not set for "spray" coagulation. Quite the contrary, the aim is for precision cutting and the energy is concentrated onto a very small area. Ours is the first reported case of an endotracheal tube fire ignited by "spray"-type coagulation cautery."

DIATHERMY

Introduction

There are two ways of using high-frequency alternating current to heat living tissue to enhance local circulation. One method employs short-wave energy; the other uses ultrahigh frequency energy. The former is designated short-wave diathermy and uses 13.56 or 27.12 MHz; the latter employs 2,450 MHz and is called microwave diathermy. Short-wave diathermy was popular in the US before about 1950 and is still used in many parts of the world. Because of the ease of application, microwave diathermy is more frequently used. With both types it takes many minutes for vasodilation to occur. A treatment may last about 30 minutes.

SHORT-WAVE DIATHERMY

Two frequencies (13.56 and 27.12 MHz) have been approved for diathermy in the US. With either frequency, two methods can be used to deliver the radio frequency energy; one uses capacitive coupling, as shown in Figure 5.13A; the other employs inductive coupling by wrapping a heavily insulated wire around the region to be heated, as shown in Figure 5.13B. Both methods are equally effective in heating tissues.

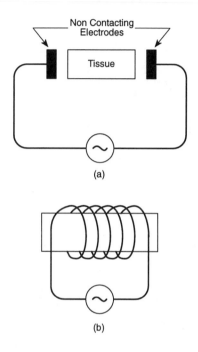

Figure 5.13. In A is shown capacities coupling and in B is illustrated inductive coupling for diathermy.

With the capacitive method, the metal electrodes are usually contained in a glass chamber (shoe) that entirely covers them and a heavily insulated cable connects them to the high-frequency (short-wave) generator. Alternately, heavily insulated pliable electrodes, usually wrapped in a towel, are placed against the skin in the region to be heated. The diathermy unit contains a timer so that the heat is applied for a selected time.

With the inductive method shown in Figure 5.13B, the high-frequency current in the heavily insulated coil surrounding the tissue induces an eddy current to produce the heating. Usually a towel is placed on the subject before the coil is applied. Sometimes a flat pancake coil is used to deposit the energy. Note that in both cases, the part of the body to be heated is wrapped with a towel to absorb perspiration that usually accompanies the heating. Without the towel, sweat accumulation could cause a hot spot on the skin due to current flow in the perspiration.

Because a towel is a good thermal insulator, vigilance must be exercised to prevent overheating.

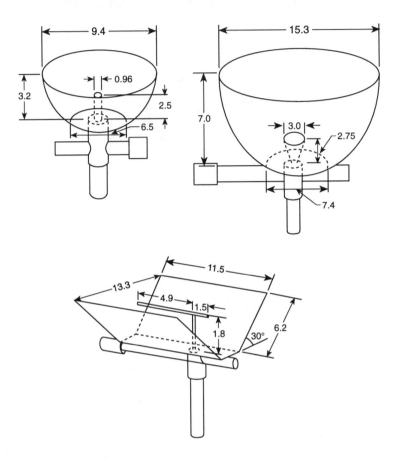

Figure 5.14. Applicators for microwave diathermy

MICROWAVE DIATHERMY

In the U.S., 2,450 MHz is allocated to diathermy; in several other countries 915 MHz is also available for this purpose. Guy et al. (1974) reported that penetration of energy is better at 915 MHz than 2,450 MHz. Because of the high frequency, the energy is delivered from several types of radiators, as shown in Figure 5.14. The energy from the microwave generator is usually delivered by a coaxial cable. Typically the maximum power available is several hundred watts.

Figure 5.14A and B illustrate hemispherical radiators which differ only in dimensions; Figure 5.14C shows the corner reflector. With each, the energy is directed into the tissue with a beam that has essentially the dimensions of the radiator. However, the field intensity across the surface of the skin is not uniform. With the hemispherical radiators, the field intensity at the center is approximately one-half of that near the perimeter. In the case of the corner reflector, the field is slightly oval, with the maximum intensity in the central area. To apply heat with microwave diathermy, the radiator is placed a specified distance from the tissue to be heated. Often a towel is placed on the skin to absorb any perspiration that accumulates. With a homogeneous medium in front of the radiator, the heating is maximum where the field strength is highest, i.e., at the skin surface. However, in a multilayered medium, such as tissues with differing electrical and thermal properties, the heating pattern is complex and heating is highest at the boundaries of subcutaneous tissues.

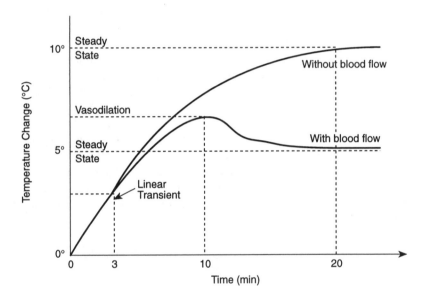

Figure 5.15. Temperature rise versus time for tissue heating without and with blood flow. (From Guy, A.W. et al. Proc. IEEE 1974, 62(1):55-75. By permission)

SUBJECT RESPONSE

With both short-wave and micro-wave diathermy, there is a pleasant sensation of warmth as blood vessels dilate in the field. Accompanying the localized vessel dilation is an increase in blood flow, an increase in cell membrane permeability and an increase in metabolic rate. It is believed that a tissue temperature of above 41°C is needed to obtain a beneficial effect. Raising tissue temperature from 34 to 40°C increases the local metabolic rate 77%. A temperature of near 45°C would produce maximum benefit; however a skin temperature of 45°C is the threshold for pain (Guy et al. 1974). During treatment, a patient will often fall asleep; therefore caution must be exercised to assure that overheating does not occur. Although heating in the tissues is nonuniform, the blood flow tends to blunt the transition between hot and cool tissues. Guy et al. (1974) illustrated this point by calculating the temperature rise without and with blood flow; Figure 5.15 shows this effect. Note that it takes many minutes for the temperature to rise and that with blood flow, the temperature rise is not as great.

DIATHERMY HAZARDS

There can be several different types of hazard associated with short-wave and microwave diathermy. Because heat is the intended therapeutic agent, excessive heating can occur. Reaching the threshold for thermal skin pain indicates that too much power is being delivered. If the patient is asleep or is sedated, this signal may not be available; therefore vigilance on the part of the therapist is essential. Because perspiration can accumulate on the skin, eddy currents can be induced and produce a skin injury. Accidental contact with an exposed conductor carrying short-wave current to capacitor electrodes or the coil encircling a body segment, can cause a skin burn. Likewise, contact with the antenna in a microwave reflector can cause a burn. Careful surveillance of patients undergoing diathermy treatments is essential. Finally the presence of a metal implant in the region exposed to radiofrequency current can produce a high temperature around the implant. Diathermy is contraindicated in such situations.

POWER-HANDLING CAPABILITY OF THE BODY, AND MAGNETIC RESONANCE IMAGING

In magnetic resonance imaging (MRI), the whole body (or part of it) is exposed to a radio-frequency field, the frequency of which is dictated by the strength of the static magnetic field. To achieve proton resonance, the frequency must be equal to 42.57 MHz per Tesla. A typical MRI with a static field of 1.5 T requires an RF field of $1.5 \times 42.57 = 63.85$ MHz. Only the magnetic field portion of radiofrequency fields is useful in magnetic resonance imaging and efforts are made to minimize radio-frequency electric fields. The radio-frequency fields used in MRI can result in induction heating in patients, somewhat analogous to induction diathermy. Fortunately, in MRI, power deposition is mostly peripheral, easing the body's task of dissipating this power to the environment.

The body is able to dissipate an electrically induced heat load. To provide perspective on this issue, it is of value to consider the heat production in a resting 70-kg human subject with a basal oxygen consumption of 250 mL/min or 4.17 mL/sec. On a typical diet, 1 mL of oxygen consumed produces 4.8 calories; therefore the number of calories produced per second is $4.8 \times 4.17 = 20.0$. The energy equivalent for 1 calorie is 4.18 joules and one joule per second is one watt. Therefore the thermal power produced by the 70-kg subject at rest is $4.18 \times 20 = 83.6$ watts. This resting power production can serve as a reference for evaluating the effect of an imposed power load. For example, walking at 2.5 mph produces 232 watts; vigorous exercise produces 522 watts. These loads raise body temperature, which is prevented from rising excessively by heat loss due to conduction, convection, radiation and the evaporation of perspiration.

Schaefer (1988) reported temperature-rise studies in unshorn, anesthetized sheep and conscious human subjects during MR imaging. In the sheep study, head and whole-body imaging were performed and temperatures were measured at many sites to identify any hot spots. The human study involved only whole-body imaging. In both cases, the weights of the subjects were about 70 kg.

Anesthetized, unshorn sheep (av. wt. 70.2 kg) were studied in an environment with an ambient temperature of $19 \pm 1 °C$ and relative humidity of 50%. Four sheep were subjected to 4 W/kg head scans for an average of 75 min (range 60 to 105 min) while cornea, vitreous humor, skin of head, tongue, jugular vein and rectal temperatures were measured intermittently with thermocouples in the absence of RF power. Similar temperatures were monitored in six animals at whole-body average power levels of 1.5, 2, or 4 W/kg for 70 to 95 min.

Surface temperatures of the eye, skin of the neck and head rose relatively rapidly in the first 20 min of imaging and then stabilized at 1 to $1.5 °C$ above the preexposure baseline level. Temperatures of the vitreous humor and tongue rose at a slightly lower rate and did not reach their highest values until near the end of the imaging period. Temperature of the jugular vein, which represents a relative measurement of the total heat load absorbed by the head, increased by $0.75 °C$ towards the end of the MR imaging period. The over-all effect of a 4 W/kg head scan on principal temperatures of the head and body core are summarized in Table 5.4. The largest increase in temperature occurred in the tongue. The temperature of the cornea and vitreous humor of the eye increased by 1.46 and $1.17 °C$, respectively. Rectal and jugular vein temperatures rose slightly during the head scan.

**Table 5.4: Effect of a 4 W/kg Head Scan on the Maximal Change
in Temperature in Anesthesized Sheep**[*]

	Cornea	Vitreous humor	Tongue	Jugular vein	Rectum
Mean	1.46	1.17	2.32	0.46	0.22
Standard error	0.13	0.3	0.87	0.05	0.008
N	4	4	4	4	4

Change in temperature (°C)

[*] from Schaefer, D.J. 1988

In the whole-body imaging, rectal and jugular vein temperatures rose approximately 1°C; while abdominal skin temperature rose as much as 7°C to temperatures approaching that of the core. Initially, abdominal skin temperature rose sharply, with rectal and jugular vein temperatures lagging by 10 to 20 min. Some skin sites on the abdomen rose by as much as 7.0°C during the scan to temperatures near that of the core. Jugular and rectal temperature rose by approximately 1.0°C, following 60 minutes of imaging. At the termination of scanning, skin temperature of the neck and abdomen fell rapidly, while the core temperature decreased more slowly.

The sheep study was also designed to assess any potential danger of RF-induced cataractogenesis, especially during a head scan. Radiofrequency-induced cataractogenesis has been produced experimentally at relatively high frequencies (greater than 2.4 GHz) and is always associated with extremely high tissue temperatures. In the present study, a 4 W/kg head scan lasting 60 min led to a maximal increase in corneal and vitreous humor temperature of 1.5 and 1.2°C, respectively. These temperature elevations are far below the thresholds necessary for any thermally induced tissue damage. Indeed, no evidence of cataract formation was found in sheep 10 weeks after exposure to whole-body or head MR imaging.

Following the sheep studies, temperature measurements were made in human subjects. Twelve adult, lightly clothed, human volunteers (average age 43.9 years and average weight 81.4 kg) were exposed for 20 min to a 4.0 W/kg whole-body, nonimaging scan at 1.5 T (64 MHz) with an ambient temperature and relative humidity of 19°C and 50% respectively. The MR imager operated in the quadrature mode (circularly polarized RF magnetic field).

Each subject was placed supine in the bore of the magnet so that xiphoid process was in the isocenter of the magnet and RF coil. Each study lasted a total of 60 min, which included a pre-RF exposure period of 20 min, and RF exposure period of 20 min, and a post-RF exposure recovery period of 20 min. As a precaution, if at any time the rise in esophageal temperature exceeded 0.7°C, the MR imaging would be aborted.

Due to technical difficulties, it was only possible to measure the esophageal temperature (T_{es}) immediately before and after the MR imaging. The mean T_{es} for the 12 subjects was 37.25°C immediately prior to the MR imaging, and increased by an average of 0.3°C following a 20-min 4 W/kg imaging. Twenty minutes after termination of the scan, T_{es} decreased by an average of 0.15°C, but did not completely recover to the pre-scan level. However, the average rise in following the MR imaging did not exceed one standard deviation of the pre-imaging T_{es}. The elevation in T_{es} following termination of the imaging is statistically significant (paired t-test: $t = 6.5$; $p < 0.005$).

With a power level of 4 W/kg, skin temperature started a sharp increase approximately 6 to 8 min following start of the imaging. Upon termination of imaging, skin temperature had increased by approximately 2°C and remained elevated for at least 20 min following the imaging. Skin sites in the abdomen and thorax area were more affected by MR imaging, compared with body areas outside the body coil. Overall, the measured xiphoid skin temperature before and after imaging indicated that mean skin temperature of the xiphoid region and abdomen increased by 2 to 3°C, while the suprapubic and forehead temperature increased by 1°C or less. Statistical assessment of these data using a paired t-test indicated that all skin temperatures, as well as T_{es}, did increase following the MR imaging. There was a slight elevation in heart rate during the MR imaging which was not statistically significant. Breathing rate was variable and apparently unaffected by MR imaging. Metabolic rate exhibited a slight elevation which was not statistically significant, immediately following the imaging.

The objective of the foregoing human study was to investigate possible thermally induced ill effects caused by MR imaging. Only a 0.3°C rise in core temperature was found for the human subjects. Although this temperature increase was statistically significant, it did not exceed one standard error of the preexposure T_{es}. True hyperthermia is defined as present when the core temperature exceeds at least one standard deviation from the normal temperature. Therefore, MR imaging at 4 W/kg does not produce hyperthermia in a 20-min scan period. Moreover, this temperature elevation is very small compared to the day-to-day circadian variation of 1 to 2°C in core temperature in adult humans. Much higher core temperatures may be safely tolerated by humans. Schaefer (1992) summarized recent body-temperature-rise studies obtained during MR imaging; Table 5.5 presents the results.

Table 5.5: Temperature Rise due to radio frequency fields.[*]

Subject & Condition	Power watts/kg	Exposure Time (minutes)	Temperature Rise °C
Human	4	20	0.3
Unshorn sheep	.2	60	0.76
Human	2.75	60	1.0
Human	2–4 (3.35 av)	30	0.1

[*] Data from Schaefer, D.J. In The Physics of MRI. Medical Physics Monograph No. 12, edited by P. Sprawls, Amer. Assoc. Physicists in Med. Amer. Inst. Physics 1992.

From the foregoing it can be concluded that there are no thermally induced ill effects associated with MR imaging as it is practiced now. The US Food and Drug Administration (FDA) has developed standards for the permissible temperature rise with MRI; a summary follows:

a) SAR < 0.4 W/kg whole body, and if the spatial peak SAR < 8 W/kg in any one gram of tissue, and if SAR < 3.2 W/kg averaged over the head, or

b) if exposure to radio-frequency magnetic fields is insufficient to produce a core temperature increase in excess of 1°C and localized heating greater than 38°C in the head, 39°C in the trunk, and 40°C in the extremities.

Note that the standard addresses total (core-body) heating as well as regional heating.

TRANSCUTANEOUS HF POWER TRANSMISSION

The foregoing dealt with HF power delivered to the whole body or a part thereof and the resulting temperature rise was measured. Other studies have been conducted in which HF power has been transmitted through the skin, the objective being to energize an implanted device such as a totally implanted artificial heart which required about 50 watts. Because it has been found that electrically stimulated skeletal muscle can be used to pump blood, research on the totally implanted artificial heart has been largely abandoned. Nonetheless the research on transcutaneous power transmission clearly showed that substantial power can be safely transmitted across the chest wall.

Figure 5.16A illustrates the method described by Schuder et al. (1961) in which a 10-watt lamp was illuminated by 350-kHz energy transmitted into the dog chest by two coils L_1 and L_2 and out of the chest by coils L_3 and L_4; the two internal coils L_3 and L_4 formed a series resonant circuit with a 250-pF capacitor. One animal carried the external oscillator in a back pack for 8 hours/day for 12 days with no adverse effect.

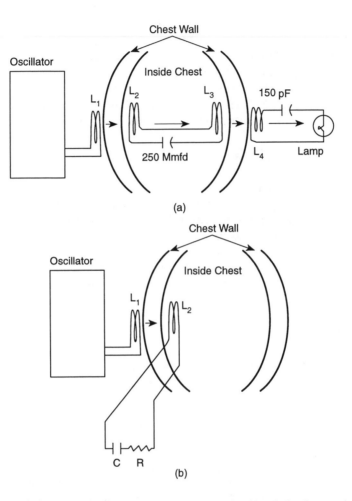

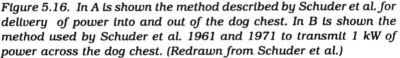

Figure 5.16. In A is shown the method described by Schuder et al. for delivery of power into and out of the dog chest. In B is shown the method used by Schuder et al. 1961 and 1971 to transmit 1 kW of power across the dog chest. (Redrawn from Schuder et al.)

In another study by Schuder et al. (1971), 1,000 watts of power at 428 kHz was delivered through the chest wall into an implanted coil, the leads from which were exteriorized and connected to a load resistor (R). The circuit was operated at resonance by the inclusion of a capacitor C, as shown in Figure 5.16B. During a one-hour period of power delivery at 1 kW, the tissue temperature at the surface of the implanted coil increased from 98°F to 103.5°F. The diameter of the implanted coil was 8 cm.

The foregoing studies demonstrate that substantial power can be transmitted though the chest wall. It is important to recognize that the power that was transmitted was confined to a relatively small region, namely that between the coils and it is only this region of tissues that was heated because the delivered power was dissipated outside of the body. Heating of living tissues dilates the small blood vessels and increases local blood flow which carries heat away. A typical blood flow for skeletal muscle at rest is about 0.15 mL/min per 100 gm. With heating this blood flow can increase manyfold, thereby providing effective heat removal.

CONTACT HF BURNS

Accidental contact with a conductor carrying high-frequency current produces a burn somewhat different than that due to low-frequency current. High-frequency current penetrates the outer layers of skin easily and high-frequency burns tend to be deep, painful and slow to heal. Having been an amateur radio operator, an engineer at a 50-kW radio station, and a researcher with electrosurgery and short-wave and microwave diathermy, the author has had experience with painful burns from all of these HF sources.

REFERENCES

AAMI (American Association for the Advancement of Medical Instrumentation), 3330 Washington Blvd., Arlington VA 22201-4598. AAMI Order Code HF18-R-2/93 Proposed 2nd edition of the American National Standard for Electrosurgical Devices.

Babbs, C.F., Bourland, J.D., Graber, G., Jones, J.T. and Schoenlein, W.E. A pressure-sensitive mat for measuring contact pressure distribution of patients lying on hospital beds. Med. Instr. Technol. 1990, 24:363–370.

Becker, C.M., Malhotra, I.V. and Hedley-White, J. The distribution of radio-frequency current and burns. Anesthesiology 1974, 38(2):106–122.

Cushing, H. and W.T. Bovie. Electro-surgery as an aid to the removal of intracranial tumors. Surg. Gynec. Obst. 1928, 47:751–784.

Foster, K.S. and Geddes, L.A. The cause of stimulation with electrosurgical current. Med. Instrum., 1986, 20(6):335–336.

Geddes, L.A. Ventricular fibrillation due to low and high-frequency electrical current. Proc. 14th AAMI Conference, 1979, 14:88.

Geddes, L.A., Tacker, W.A. and Cabler, P. A new electrical hazard associated with the electrocautery. Med. Instrum., 1975;9(2):112–113.

Geddes, L.A., Pearce, J.A., Bourland, J.D., Silva, L.F. and De Witt, D.P. The thermal properties of gelled and metal-foil electrosurgical dispersive electrodes. Proc. 14th AAMI Conference, 1979, 14:90.

Geddes, L.A. Ventricular fibrillation due to low and high-frequency electrical current. Proc. 14th AAMI Conference, 1979, 14:88.

Geddes, L.A., Pearce, J.A., Bourland, J.D. and Silva, L.F. Thermal properties of dry metal-foil dispersive electrodes. Clinical Engineering, 1980, 5:13–18.

Geddes, L.A. and Moore, C. Stimulation with electrosurgical current. Physics and Eng. Sci. in Med., 1990, 13(2):63–66. (Australia)

Geddes, L.A., Silva, L.F., DeWitt, D.P. and Pearce, J.A. What's new in electrosurgical instrumentation? Med. Instr. 1977, 11(6):385–389.

Guy, A.W., Lehmann, J.F. and Stonerbridge, J.B. Therapeutic applications of electromagnetic power. Proc. IEEE 1974, 62(1):55–75.

Hungerbuhler, R.F. and Swope, J.P. Ventricular fibrillation associated with the use of electrocautery. JAMA 1974, 230(3):431–435.

Irnich, W. How to avoid surface burns during electrosurgery. Med. Instr. 1986, 20(6):320–326.

Lawson, N.W., Mills, N.L. and Ochser, J.L. Occipital alopecia following cardiopulmonary bypass. J. Thor. Cardiovasc. Surg. 1976, 71:342–347.

Pearce, J.A., Geddes, L.A., Smith, J., Bourland, J.D., Silva, L. and Jones, J.T. The performance of dry and gelled electrosurgical electrodes. Proc. 31st ACEMB, paper 25.2 (Atlanta, GA), 1978.

Pearce, J.A., Geddes, L.A., Bourland, J.D., Silva, L.F. and De Witt, D.P. The effect of perimeter/area ratio on the performance of electrosurgical dispersive electrodes. Proc. 14th AAMI Conference, 1979, 14:196.

Pearce, J.A. Electrosurgery. London 1986, Chapman & Hall, 258 pp.

Pearce, J.A. and Geddes, L.A. The characteristics of capacitive electrosurgical dispersive electrodes. Proc. 15th AAMI Conference, 1980, 15:162.

Pearce, J.A., Geddes, L.A., Bourland, J.D. and Silva, L.F. The thermal behavior of electrolyte-coated metal-foil dispersive electrodes. Med. Instrum., 1979, 3(5):298–300.

Pearce, J.A., Geddes, L.A., Van Vleet, J., Foster, K. and Allen, J. Skin burns from electrosurgical electrodes. Med. Instrum., 1983, 17:225–231.

Prentiss, G.W.H., Bethard, W.F., Boatwright, D.E. and Pennington, R.D. Massive adductor muscle contraction in transurethral surgery. J. Urol. 1965, 93:263–271.

Rita, L. and Seleny, F. Endotracheal tube ignition during laryngeal surgery with resectoscope. Anesthesiology 1982, 56:60–61.

Schaefer, D.J. Safety aspects of magnetic resonance imaging. Wehrli, F.W., Shaw, D. and Kneeland, B. eds. New York, 1988 VCH Publishers.

Schaefer, D.J. Bioeffects of MRI and Patient Safety. In The Physics of MRI, 1992 AAPM Summer School Proceedings. Amer. Assoc. Physicists in Med. Amer. Inst. Physics.

Schuder, J.C., Stephenson, H.E. and Townsend, J.F. Energy transfer into a closed chest by means of stationary coupling coils and a portable high-power oscillator. Trans. ASAIO, 1961, 7:327–331.

Schuder, J.C., Gold, J.H. and Stephenson, H.E. An inductively coupled RF system for transmission of 1 kW of power through the skin. IEEE Trans. Biomed. Eng. 1971, July:265–273.

Simpson, J. and Wolf, G.L. Endotracheal tube fire ignited by pharyngeal electrocautery. Anesthesiology 1986, 63:76–77.

Tucker, R.D., Schmitt, O.H., Silvert, C.E. and Silvis, S.E. Demodulated low
    frequency currents from electrosurgical procedures. Surg. Gyn. Obst. 1984,
    159:39–43.
Ward, G.E. An efficient method of hemostasis without suture. Med. J. & Record
    1925, 121:470.

Chapter 6
ELECTRICAL PROPERTIES OF LIVING TISSUES

INTRODUCTION

A living subject consists of tissues and fluids with differing conducting properties. Current injected into a subject will seek the tissues and fluids with the lowest resistivities. Resistivity is the fundamental property that describes the opposition to current flow, and is not dependent on the amount of material present. Resistance is the opposition to current flow for a given geometric shape of the material. For a conductor of length L (cm) and a uniform cross-sectional area A (sq. cm), the resistance R in ohms is equal to $\rho L/A$, where ρ is the resistivity in ohm-cm. Resistivity values are available for many, but not all tissues. For the values that are available, not all authors reported complete details of measurement. Moreover, the values given are for different frequencies. In the tables that follow, resistivity values are given for inductorium (induction-coil) current which is typically a short-duration (100 μsec) pulse of exponentially decaying current with a repetition rate of about 100/sec, sinusoidal current and direct-current pulses.

IMPEDANCE LOCUS PLOT

Because cell membranes have dielectric (insulating) properties, a tissue specimen can exhibit a small capacitive (C) component, depending on the frequency used for measurement. The resistivity for direct current represents current flowing mainly in the fluid surrounding the cells. The resistivity for high-frequency current represents current flow in the extracellular fluid as well as through the intracellular fluid. Figure 6.1A illustrates current flow around a cell in an electrolyte for direct current; Figure 6.1B shows the path for high-frequency current. Figure 6.1C illustrates the impedance-frequency relationship for a typical specimen. To demonstrate the reactive component, it is convenient to plot the reactance ($X = 1/2\pi fC$) versus the resistance (R) exhibited by the specimen; the result is a diagram that resembles the sector of a circle, as shown in Figure 6.1D. Such a diagram is called an impedance - locus plot. Observe that the impedance (Z) of the specimen is represented by the line from 0,0 to any point on the diagram and points on the sector are identified by frequency. It is obvious that there is a frequency for the maximum phase angle (θ_{max}).

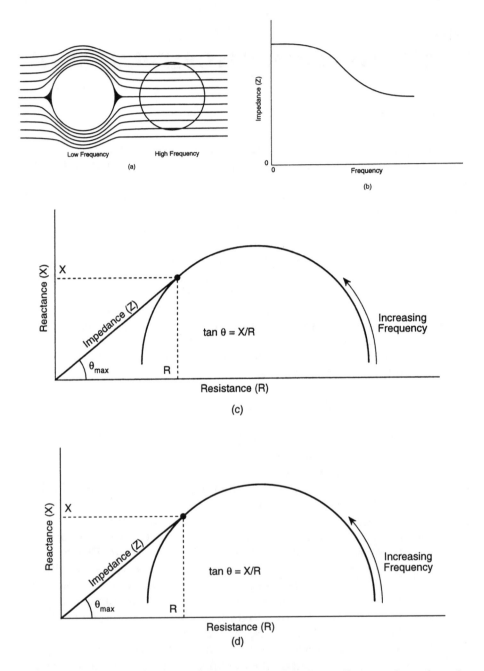

Figure 6.1. Idealized current flow associated with a cell in an electrolyte. In A is shown direct current flow around the cell, and in B is shown the path for high-frequency current. In C is shown a typical impedance–frequency relationship, and D is an impedance–locus plot.

There are not many reports that describe the capacitance of cell membranes and the resistivity of the fluid within a cell (cytoplasm). The capacitance is normalized to $\mu F/cm^2$ and the resistivity is expressed in ohm-cm. Table 6.1 lists typical values for a few different types of cells. It is interesting to observe that the capacitance (C) of typical cell membranes is about $1\mu F/cm^2$.

Table 6.1: Membrane Capacitance and Cytoplasm Resistivity

Cell Type	Membrane Capacitance $\mu F/cm^2$	Cytoplasm resistivity Ω-cm	Investigator and Year
Red cell (man)	0.8	--	Fricke 1931
Red cell (turtle)	0.8	140	Fricke & Curtis 1934
Leukocyte (rabbit)	1.0	140	Fricke & Curtis 1935
Lymphocyte (turtle)	0.8	140	Fricke & Curtis 1935
Sartorius muscle (frog)	1.5	250	Bozler & Cole 1935
Sciatic nerve (cat)	0.65	720	Cole & Curtis 1936
Sciatic nerve (frog)	0.55	560	Cole & Curtis 1936
Egg (frog)	2.0	570	Cole & Guttmann 1942

RESISTIVITY MEASUREMENT

Typically, resistivity is measured with the bipolar or tetrapolar method. In the former case, low-impedance electrodes made of platinum-black are used along with a frequency that is high enough so that the electrode-electrolyte impedance is negligible. Direct-current resistivity cannot be measured accurately with bipolar electrodes. With the tetrapolar-electrode method, errors due to the impedance of the electrode-electrolyte interfaces are eliminated, because two outer electrodes are used to inject current and two inner electrodes are used to measure the potential with a device that has a very high input impedance. Resistivity measurements down to zero Hz (dc) can be made easily with the tetrapolar method.

EFFECT OF TEMPERATURE

Because living tissues are largely made up of electrolytes, temperature is an important factor. Electrolytes decrease their resistivity with an increase in temperature. A typical value is a decrease of a few percent for a 1°C rise in temperature. When viewing the values for tissue resistivity, this fact should be kept in mind.

TISSUE RESISTIVITY

Typical values for tissue resistivity are presented in the accompanying tables. Many of the values were derived from the review presented by Geddes and Baker (1967); supplemental data are from more recent papers.

Table 6.2: Body Fluids

Substance	Resistivity (Ω-cm)	Frequency	Temp. (°C)	Electrodes	Reference
C.S.F.					
Human	64.6(64.0-6-5.2)	1–30 kHz	24.5	Not given	Radvan-Ziemnowicz, 1964
Cat	65.7(65.5-6-6.1)	1–30 kHz	24.5	Not given	
Rabbit	55.9 avg (51–62)	1 kHz	39	2	Crile, 1922
Bile					
Cow-pig	60	Audio	37	2	Osswald, 1937
	78	Audio	20	2	
	59	50 MHz	37	2	
	76	50 MHz	20	2	
Rabbit	66.2 avg (61–72)	1kHz	39	2	Crile, 1922
Amniotic fluid					
Sheep	65	1 kHz	25	2	Unpublished
	49	1 kHz	37.5	2	
Urine					
Cow-pig	30	Audio	37	2	Osswald, 1937
	39	Audio	20	2	
Milk					
Cow (whole)	215	1 kHz	24	2	Author
Cow (whole)	170	1 kHz	36	2	
Cow (skim)	210	1 kHz	24	2	
Cos (skim)	175	1 kHz	36	2	

BODY FLUIDS

Table 6.2 presents resistivity values for various body fluids (except blood). Note that at body temperature, urine has the lowest resistivity, with amniotic fluid being slightly higher. The other fluids have resistivities that are approximately the same as that of plasma.

BLOOD

Values for the resistivity of blood at body temperature for several species are presented in Table 6.3. Note that the resistivity depends on packed-cell volume (H), which is the percentage of cells in a sample. The values for H = 0 represents plasma.

Table 6.3: Resistivity of Blood at Body Temperature

Species	Exponential Expression	Frequency (kHz)	Reference
Human	$62.9e^{0.0195H}$	1	Rosenthal & Tobias (1948)
Human	$53.2e^{0.022H}$	25	Geddes & Sadler (1973)
Dog	$56.6e^{0.022H}$	100	Kinnen et al. (1964)
Dog	$53.7e^{0.025H}$	25	
Cow	$54.2e^{0.020H}$	25	Geddes & Sadler (1973)
Horse	$57.0e^{0.024H}$	25	
Sheep	$55.6e^{0.020H}$	25	
Goat	$58.3e^{0.015H}$	25	
Cat	$53.9e^{0.023H}$	25	Geddes (unpublished)
Monkey	$53.7e^{0.019H}$	25	
Baboon	$57.5e^{0.019H}$	25	

It is important to recognize that blood is a suspension of cells which precipitate with the passage of time. The precipitation (sedimentation) rate depends on the species and also on the presence of inflammation. When blood sedimentates, the bottom layer is composed of red blood cells. Above it is a thin layer of platelets and white blood cells and is called the buffy coat. By definition, the volume of cells in a sample of blood is the packed-cell volume (PCV). The percentage of red-blood cells is known as the hematocrit. In the normal subject, because the buffy coat is so thin, the packed-cell volume is almost the same as the hematocrit. A typical value for hematocrit is 40%.

EFFECT OF FLOW

There is a considerable variety of shapes for red cells, ranging from biconcave disks to ellipsoids. As would be expected, the orientation of cells changes with flow velocity. Therefore the resistivity of a uniform stationary suspension of blood cells is slightly higher than the value measured when the cells are in a flowing stream which causes axial accumulation of cells. Figure 6.2 illustrates the resistivity of static and flowing bovine blood versus hematocrit. Observe that the decrease in resistivity is larger for a higher hematocrit. However, the resistivity of plasma is independent of flow velocity.

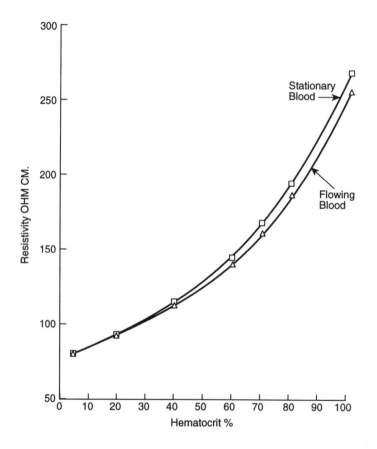

Figure 6.2. Resistivity of bovine blood versus hematocrit for stationary and flowing blood. Note that the decrease in resistivity is greater for higher hematocrit. (Drawn from data provided by Sigman 1937).

There has been considerable investigation of the decrease in resistivity with increasing flow velocity. Among those who have reported on the resistivity-rheological phenomenon are Velick and Gorlin (1940), Moskalenko and Naumenko (1959), Baker and Mistry (1981), Liebman et al. (1962-63), Kanai et al. (1976) and Sakamoto and Kanai (1979).

EFFECT OF ELECTRIC FIELD

When blood is exposed to a high electric field, the dielectric of the cell membranes breaks down and the resistivity decreases. There is some information on the reduction in resistivity of blood exposed to high electric fields. Tacker et al (1982) measured the resistivity of blood samples exposed to single 30-msec rectangular-wave shocks with field strengths ranging from 30 to 403 V/cm. Packed-cell volumes ranging from 0 (plasma) to 98% were tested. Plasma exhibited virtually no change in resistivity; the higher packed-cell volume samples exhibited a decrease in resistivity with increasing field strength. The resistivity of the 90% packed-cell volume sample decreased from 2,040 to 1,509 ohm-cm for field strengths ranging from 30 to 403 V/cm.

PHYSIOLOGICAL SOLUTIONS

Several types of electrolytic solutions are used in biomedical studies, the most common being dilute sodium chloride (saline). For warm-blooded species, 0.9% is iso-osmotic; for cold-blooded animals, about 0.6% is iso-osmotic. Iso-osmotic means that cells placed in such a solution neither swell nor shrink. Table 6.4 gives resistivity values for commonly used solutions.

Table 6.4: Resistivity of Physiological Solutions

		ρ (Ω-cm)	T° C
Saline	0.6%	98.5	21
		79.1	37
	0.9%	69.5	21
		58.3	37
	2%	37.4	21
		30.9	37
Tyrodes		52	37
3M KCl		4.28	20
		3.25	37.5

EFFECT OF CONCENTRATION ON RESISTIVITY

As stated previously, the resistivity (ρ) of an electrolyte decreases with an increase in temperature. It also decreases with an increase in the concentration (C) of the solute. For dilute solutions $\rho = k/C^\alpha$, where k and α are constants and C is the concentration of the dissolved salt. Figure

6.3 presents the relationship between resistivity (ρ in ohm-cm) and the concentration (C) of NaCl (in gm/L) at 20°C and 37°C.

Table 6.4 lists resistivity values for sodium chloride and potassium chloride solutions at various temperatures. At 37°C, the resistivity (ρ) of a dilute saline solution is $411.5/C^{0.8975}$, where C is in gm/L. Tyrode's is a modified saline solution. Three-molar (3M) or 22% potassium chloride is used to fill micropipet electrodes and is also used in salt bridges.

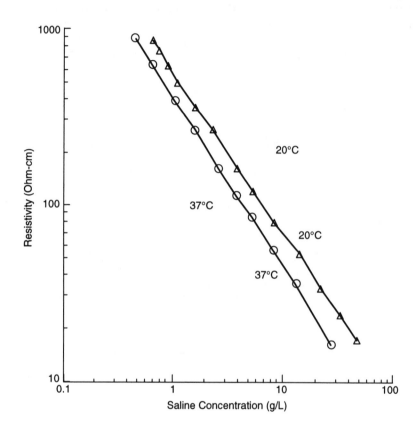

Figure 6.3. Relationship between resistivity and concentration of saline at 20°C and 37°C.

ELECTRODE PASTES

Many different electrolytic preparations have been used to establish electrical contact between a metal electrode and the skin. The low-resistivity preparations are used when current is delivered, as for example in stimulation, cardioversion and ventricular defibrillation. High-resistivity preparations are used for recording bioelectric events, such as the ECG. Such preparations resemble hand cream and replaced the older low-resistivity pastes because of the higher input impedance of modern bioelectric recorders. Table 6.5 presents a compilation of the resistivity of compounds used for recording bioelectric events. Only the low-resistivity preparations should be used with defibrillation electrodes.

Table 6.5: Resistivity of Electrode Preparations

Preparation and Supplier	Resistivity[a] (Ω-cm)
Redux Electrode Paste Sanborn Div., Hewlett Packard, Waltham, Mass.	9.4
Electrode Cream EC-2 Grass Instrument Co., Quincy, Mass.	30
Cambridge Electrode Jelly Cambridge Instrument Co., Ossining, NY	10.4
Beckman-Offner Paste Offner Division, Beckman Instrument Co, Chicago, Ill.	5.9
EKG-Sol Burton, Parsons & Co., Washington, DC	200
Burdick Electrode Jelly Burdick Co., Milton, Wis.	10
Cardiopan Leichti, Berne, Switzerland	120
Cardette Electrode Jelly Newmark Instrument Co., Croydon, Surrey, England	313
Electrode Jelly Smith and Nephew Res. Ltd., Harlow, Essex, England	118
Cardioluxe Electrode Jelly Philips Electrical Ltd., Balham, London, England	84
Electrode Jelly Data Display Ltd., Liverpool, England	196
K-Y Lubricating Jelly Johnson & Johnson, Slough, Buckinghamshire, England	323

[a] At room temperature.

MUSCLE

There are three types of muscle: skeletal, cardiac and smooth. The first moves the bony levers of the body; the second pumps blood and the third constitutes the muscular layers of many internal organs and blood vessels. Although there are data for skeletal and cardiac muscle, there are no data for discrete specimens of smooth muscle. However, values are available for organs rich in smooth muscle.

Skeletal muscle consists of bundles of long contractile fibers; therefore the resistivity depends on the direction of current flow. The resistivity for longitudinal current is lower than for transverse current. Table 6.6 presents a compilation of data for skeletal muscle.

Table 6.6: Skeletal Muscle

Species	Resistivity (Ω-cm)	Frequency	Temp. (°C)	Elec-trodes	Reference	Remarks
Human	245	d.c.	37	4	Burger, 1960–61	Longitudinal
	240	100–1000Hz	37	4	Burger, 1960–61	Longitudinal
	675	20–1000Hz	37	4	Burger, 1960–61	Transverse
	1800	20–1000Hz	37	4	Burger, 1960–61	Transverse
	1750	20–1000Hz	37	4	Burger, 1960–61	Random orientation
	125	20–1000Hz	37	4	Burger, 1960–61	Longitudinal
Rabbit	1243	Inductorium	18	2	Galeotti, 1902	Transverse current
	548	Inductorium	12	2	Galeotti, 1902	Longitudinal current
Cow & Horse	300	20–5000Hz	33	4	Burger, 1960–61	Longitudinal
	700	20–5000Hz	33	4	Burger, 1960–61	Transverse
	550	20–5000Hz	33	4	Burger, 1960–61	Random orientation
Dog	408	Inductorium	12	2	Galeotti, 1902	Longitudinal current
Dog-adult	1072	Inductorium	12	2	Galeotti, 1902	Transverse current
Turtle	1350av	Inductorium	18	2	Galeotti, 1902	Transverse current
	740av	Inductorium	18	2	Galeotti, 1902	Longitudinal current

Cardiac muscle constitutes the heart chambers. Like skeletal muscle fibers, cardiac muscle fibers are long with respect to their diameters; therefore the longitudinal resistivity is expected to be lower than the transverse resistivity. However, few investigators have reported resistivity values for the two current pathways. Table 6.7 presents values for the resistivity of cardiac muscle.

Table 6.7: Cardiac Muscle

Species	Resistivity (Ω-cm)	Frequency	Temp (°C)	Electrodes	Reference	Remarks
Human	563	d.c. pulses 0.1 sec	Body	4	Rush, 1963	Transverse to fibers
	252	d.c. pulses 0.1 sec	Body	4	Rush, 1963	Parallel to fibers
	965 av	10Hz	Body	2	Schwan, 1956–57	Anesthesized subject
	1250	10Hz	Body	2	Schwan, 1956–57	Anesthesized subject
	925	100Hz	Body	2	Schwan, 1955	Anesthesized subject
	1150	100Hz	Body	2	Schwan, 1955	Anesthesized subject
	215 av (207–224)	1kHz	Body	2	Kaufman, 1943	Anesthesized subject
Dog	875 av (750–1000)	1kHz	Body	2	Schwan, 1956	Anesthesized animal
	825 av (700–950)	1kHz	Body	2	Schwan, 1955	Anesthesized animal
	845	1kHz	Body	2	Schwan, 1956–57	Anesthesized animal
	600	10kHz	Body	2	Schwan, 1956–57	Anesthesized animal
	825 av (700–950)	10kHz	Body	2	Schwan, 1955	Anesthesized animal
	456	100kHz	Approx. body	2	Kinnen, 1964	Left ventricle

LUNG

Lung tissue, being constituted largely by air sacs (alveoli), is expected to exhibit a higher resistivity than fluid-rich organs. The resistivity of lung reflects the air content. The first few entries in Table 6.8 show that the resistivity is high for inflated lungs and lower for deflated lungs.

Table 6.8: Lung[*]

Spe-cies	Resistivity (Ω-cm)	Frequency	Elec-trodes	Reference	Remarks
Dog	2390	d.c.pulses 0.1 sec	4	Rush, 1963	Peak inflation -anesth.animal
	1950	d.c.pulses 0.1 sec	4	Rush, 1963	Max. deflation -anesth.animal
Dog	1530	100kHz	2	Kinnen, 1964	Full inspiration
	1345–2100	100kHz	2	Kinnen, 1964	Mid inspiration
	1220	100kHz	2	Kinnen, 1964	Complete expiration
Dog	744–766	1kHz	2	Kaufman, 1943	End inspiration -anesth.animal
	1227–1367	1kHz	2	Kaufman, 1943	Super inflation -anesth. animal
	401	1kHz	2	Kaufman, 1943	Deflation -anesthetized
Dog	2170	d.c.pulses 0.1 sec	4	Rush, 1963	Average -anesth.animal
	1120	10Hz	2	Schwan, 1956–57	Anesthesized animal
	900–1600	10Hz	2	Schwan, 1955	Anesthesized animal
	1090	100Hz	2	Schwan, 1956–57	Anesthetized animal
	800–1500	100Hz	2	Schwan, 1955	Anesthesized animal
	1100	10–100k-Hz	2	Schwan, 1965	Anesthesized animal
Dog	800–1300	1kHz	2	Schwan, 1955	Anesthesized animal
	1040	1kHz	2	Schwan, 1956	Anesthesized animal
	750–1000	1kHz	2	Schwan, 1956	Anesthesized animal
	950	10kHz	2	Schwan, 1956–57	Anesthesized animal

[*] All measurements at body temperature.

KIDNEY

The resistivity of the kidney has been of interest because of the urine that the kidney produces. Urine is the lowest-resistivity fluid in the body. Therefore, because of its urine content, kidney resistivity would be expected to be low, as shown in Table 6.9.

Table 6.9: Kidney

Species	Resistivity (Ω-cm)	Frequency	Temp. (°C)	Elec-trodes	Reference	Remarks
Human	126	1MHz	Near room	2	Heming-way, 1932	2–3 hr after death
	94av (81–104)	200–900 MMz	27	2	Schwan, 1955	Autopsy material
Cow-pig	111	Audio	37	2	Osswald, 1937	
	119	50–100 MHz	37	2	Osswald, 1937	
	143	Audio	20	2	Osswald, 1937	
	147	50–100 MHz	20	2	Osswald, 1937	
	204	25MHz	20	2	Osswald, 1937	
Dog	272	Inductorium	38	2	Galeotti, 1902	3 min after extirpa-tion
	600	100kHz	Body	2	Kinnen, 1964	

SPLEEN

Among its many functions, the spleen is a reservoir for red-blood cells. Because the resistivity of packed cells is high, the spleen is expected to exhibit a somewhat higher resistivity than that of other smooth-muscle organs. Table 6.10 presents resistivity values for spleen.

Table 6.10: Spleen

Species	Resistivity (Ω-cm)	Frequency	Temp. (°C)	Elect- rodes	Reference	Remarks
Human	256	1MHz	Near room	2	Hemingway 1932	2–3 hrs after death
Dog	885	Inductorium	38	2	Galeotti 1902	Freshly extirpated
	1053	Inductorium	24	2	Galeotti 1902	Freshly extirpated
Dog- adult	1010av	Inductorium	24	2	Galeotti 1902	
	1040av	Inductorium	18	2	Galeotti 1902	
	1196av	Inductorium	12	2	Galeotti 1902	
Dog	1178	Inductorium	12	2	Galeotti 1902	Freshly extirpated
Cow- pig	715av	Audio	37	2	Osswald 1937	
	137av	50MHz	37	2	Osswald 1937	
	120av	100MHz	37	2	Osswald 1937	
	833	Audio	20	2	Osswald 1937	
	175	25MHz	20	2	Osswald 1937	
	156	50MHz	20	2	Osswald 1937	
	147	100MHz	20	2	Osswald 1937	

NERVE TISSUE

Resistivity measurements have been made on nerve tissue for a variety of reasons. In electroshock therapy, electrodes are placed on the head and prediction of the current path requires a knowledge of the resistivity of the brain, scalp and skull. Electrical accident studies also need information on the conducting properties of the nervous system. Finally, fundamental studies have been done to determine the longitudinal and transverse resistivity of nerve fibers. Table 6.11 presents typical values for the resistivity of neural tissue. As expected, the transverse resistivity of bundles of fibers is higher than the longitudinal values.

Table 6.11: Nerve Tissue

Species and tissue	Resistivity (Ω-cm)	Frequency	Temp (°C)	Electrodes	Reference	Remarks
Brain	588av	Audio	37	2		
	222	25MHz	37	2	Osswald 1937	
	196av	50MHz	37	2		
Rabbit cerebellum	570av (521–725)	1kHz	39	2	Crile 1922	
	730av (610–855)	1kHz	39	2		
Cat-internal capsule	800av	20Hz–20kHz	Body	3	Nicholson 1965	Transverse to fibers - anesth.
	85av	20Hz–20kHz	Body	3		Along fibers - anesth.
Rabbit cortex	321	5Hz	Body	4	Ranck 1963	Anesthetized
	230	5kHz	Body	4		
	208±6	1kHz	Body	3	Van Harreveld 1963	Anesthetized
Rabbit white matter	957 (approx)	1kHz	Body	3		
Rabbit cerebellum	662–794	1kHz	39	2		
	505–621	1kHz	39	2		
Rabbit spinal cord	576av (386–863)	1kHz	39	2	Crile 1922	
Rabbit cerebral (gray)	438av	1kHz	39	2		
Rabbit cerebral (white)	746av	1kHz	39	2		
Cat cortex	222±9	Square pulses 0.3–0.7 msec	37	4	Freygang 1955	Anesthetized animal
Cat white matter	344 (approx)	0.3–0.7 msec	37	4		
Cat spinal cord	138–212	5–10Hz	Body	4	Ranck 1965	Longitudinal
	1211	5–10Hz	Body	4		Transverse

LIVER

Liver is a highly vascularized tissue that performs many biochemical functions. It receives arterial blood and venous blood from the portal system and drains its output into the inferior vena cava. Table 6.12 presents resistivity values for liver, indicating that the range of variability is quite large.

Table 6.12: Liver

Species	Resistivity (Ω-cm)	Frequency	Temp (°C)	Elec-trodes	Reference	Remarks
Human	128av (92–170)	200–900 MHz	27	2	Schwan 1953	Autopsy material
Cow-pig	833av	Audio	37	2	Osswald 1937	
	192av	25MHz	37	2	Osswald 1937	
	182av	50MHz	37	2	Osswald 1937	
	164av	100MHz	37	2	Osswald 1937	
Dog	700	d.c. pulses 0.1 sec	Body	4	Rush 1963	Anesthetized
	1100av (1000–1200)	10Hz	Body	2	Schwan 1955	Anesthetized
	840	10Hz	Body	2	Schwan 1956–57	Anesthetized
	800	100Hz	Body	2	Schwan 1956–57	Anesthetized
	925av (850–1000)	100Hz	Body	2	Schwan 1955	Anesthetized
	900	10Hz–100kHz	Body	2	Schwan 1965	
	900av (800–1000)	1kHz	Body	2	Schwan 1955	Anesthetized
	765	1kHz	Body	2	Schwan 1956–57	Anesthetized
	875av (750–1000)	1kHz	Body	2	Schwan 1956	Anesthetized
	589av (506–672)	1kHz	Body	2	Kaufman 1943	Anesthetized
	685	10kHz	Body	2	Schwan 1956–57	Anesthetized
	775av (700–850)	10kHz	Body	2	Schwan 1955	Anesthetized
	600av (300–900)	100kHz	Body	2	Kinnen 1964	Anesthetized
Rabbit	1235av (990–1639)	1kHz	39	2	Crile 1922	Freshly extirpated
Guinea pig	2380	1kHz	25	2		
	464	200kHz	25	2		
	317	800kHz	25	2	Philipson 1920	
	260	2MHz	25	2		
	225	3.5MHz	25	2		

PANCREAS

There is a paucity of information on the resistivity of pancreatic tissue; what is available is shown in Table 6.13.

Table 6.13: Pancreas

Species	Resistivity (Ω-cm)	Frequency	Temp (°C)	Electrodes	Reference
Cow-pig	770av	Audio	37	2	Osswald 1937
	185av	25–100MHz	37	2	Osswald 1937
	625av	Audio	20	2	Osswald 1937
	250av	25–100MHz	20	2	Osswald 1937

Table 6.14: Fat

Species	Resistivity (Ω-cm)	Frequency	Temp (°C)	Electrodes	Reference	Remarks
Human	2180	1MHz	Freshly excised	2	Hemingway	Between body and room temp.
	1500–5000	200MHz	27	2	Schwan 1953	Autopsy material
	1300–4000	400MHz	27	2	Schwan 1953	Autopsy material
	1100–3500	900MHz	27	2	Schwan 1953	Autopsy material
Dog	1500–3000	10Hz–100-kHz	body	2	Schwan 1965	
	2500	d.c. pulses 0.1 sec	body	4	Rush 1963	Anesthetized animals
	2006av (1808–2205)	1kHz	Body	2	Kaufman 1943	Anesthetized animals
	1000–3000	100kHz	Body	2	Kinnen 1964	
	1500–5000	1kHz	Body	2	Schwan 1956–57	Anesthetized
	3000 av (1500–5000)	1kHz	Body	2	Schwan 1956	Anesthetized
Cow-pig	2500av	Audio	37	2	Osswald 1937	
	2000av	25–100M-Hz	37	2	Osswald 1937	
	3850av	Audio	20	2	Osswald 1937	
	2780av	25–100M-Hz		2	Osswald 1937	

FAT

Fatty tissue is lightly vascularized with a low electrolyte content, and consequently resistivity values are expected to be high. Table 6.14 summarizes the published literature, revealing a range from 1,000 to 5,000 ohm-cm, with a typical range of 2,000 to 3,000 ohm-cm.

BONE

Bone is a complex tissue consisting of cortical bone and marrow, the site of many cellular elements. Moreover, the composition of bone varies throughout the body; therefore the resistivity values are expected to reflect this situation. Table 6.15 lists bone resistivity values.

Table 6.15: Bone

Species	Resistivity (Ω-cm)	Frequency	Temp (°C)	Electrodes	Reference	Remarks
Human (thorax)	16,000	Low	Not given	Not given	Lepeschkin 1951	(ECG Spectrum)
Human	1800	1MHz	Freshly excised	2	Hemingway 1932	Between body and room temp.
Cow-pig	4550av	Audio	37	2	Osswald 1937	
	3700av	25–100MHz		2	Osswald 1937	
	6250av	Audio	20	2	Osswald 1937	
	5000av	25–100MHz	20	2	Osswald 1937	

BODY SEGMENTS

Average resistivity values for body segments have been obtained by a few investigators. Such data, along with segment dimensions, allow for an estimate of the magnitude of current that will flow for electrodes applied at different body sites. Table 6.16 presents the available data.

Table 6.16: Body Segments

Species	Resistivity (Ω-cm)	Frequency	Temp (°C)	Electrodes	Reference	Remarks
Human arm	160	d.c.pulses 0.1 sec	Body	4	Rush 1963	Corrected for bone and fat
Human forearm	470	d.c.	Body	4	Burger 1943	Transverse

	230	d.c.	Body	4	Burger 1943	Longitudinal
	330	d.c.	Body	4	Burger 1943	Geometric mean
Fingers & Hand	280	d.c.	Body	4	Burger 1943	
Finger	235	d.c.	Body	4	Burger 1943	Current along finger
Neck	280	d.c.	Body	4	Burger 1943	
Trunk	415	d.c.	Body	4	Burger 1943	Along axis of body
Head	840	d.c.	Body	4	Burger 1943	Trans-temp-oral
Head (Scalp)	230	d.c.	Body	4	Burger 1943	Closely spaced electrodes
Thorax	455	d.c.	Body	4	Burger 1943	Maximum inspiration
	375	d.c.	Body	4	Burger 1943	Maximum expiration
	463	d.c.pulses 0.1 sec	Body	4	Rush 1963	
Dog thorax	445	d.c.pulses 0.1 sec	Body	4	Rush 1963	Intact thorax
	281	d.c.pulses 0.1 sec	Body	4	Rush 1963	Shell-less heart and lungs

CURRENT-DENSITY AND CELL-MEMBRANE BREAKDOWN

As described earlier, a living cell resembles a thin insulating pouch containing an electrolyte and bathed in extracellular electrolytes. If sufficient current is applied so that the voltage gradient across the cell membrane exceeds a critical value in volts/cm, the insulating property of the membrane breaks down. In the section dealing with blood resistivity, it was shown that above a critical voltage gradient, the resistivity of a suspension of red cells decreases. There is some information on dielectric breakdown of other cells. The current density is the electric field divided by the resistivity.

DeGaravilla et al. (1981) reported the effect of 30-msec rectangular-wave pulses on the resistivity of cardiac muscle. Fifteen pulses, producing an electric field ranging from 33 to 405 V/cm were delivered. In general the resistivity decreased with successive pulses with higher field strengths. Tacker et al. (1984) measured the decrease in resistivity of skeletal muscle, skin and fat exposed to 30-msec rectangular pulses with field strengths of 50–400 V/cm. The decrease in resistivity was most for skeletal muscle, less for skin and least for fat.

There is limited information on the breakdown of membranes of single cells exposed to high-intensity field strengths. Coster and Zimmerman (1975) used transmembrane electrodes to study the breakdown of membranes of valonia with 0.5–1 msec pulses. They found that the breakdown was local and recovery occurred in less than 4 sec. Guager and Bentrup (1979) investigated membrane breakdown in brown alga using pulses of 1–1,760 μsec in duration and field strengths of 50–400 V/cm. They too found that the breakdown was local and recovery occurred in less than 3 sec.

BODY IMPEDANCE

In 1984 the Commission Electro-technique International (Geneva) published a report giving the impedances between body sites for large-area contact. The impedance values were expressed as a percent of the hand-to-hand impedance (100%). However, the Commission stated: "The value of the initial resistance of the human body for a current path hand-to-hand or hand-to-foot and large contract areas can be taken as 500 ohms for the 5% percentile rank." Figure 6.4 is reproduced from the 1984 report. Note that the hand-to-hand path represents 100%, which corresponds to 500 ohms for a large contact area. The numbers in the circles represent the percent of the hand-to-hand impedance for the path identified. The numbers in the brackets within the circles represent the percentages when both hands are joined and current flows between both hands and the site indicated.

Although it may be useful to assign a value to body impedance for medium-voltage, the nature of the electrode-subject circuit is quite different and time-varying for higher voltages. The dynamic nature of the electrode-subject circuit was investigated by Sances et al. (1981) in two series of experiments using hogs. In both studies, a large (40×60 cm) steel plate was placed under one hindquarter and in one study a 5-cm length of #2ACSR wire was brought into contact with the skin of the opposite hindquarter. In the other study metal disks (0.5, 1, 2.5 and 5 cm) were applied to the skin with 0.1 kg/cm^2 pressure. In both cases, current was measured with the application of voltages up to 14,400. With 10–80 volts applied to the #2ACSR conductor, it took up to 200 seconds for the current to reach maximum values. The current buildup was coincident with muscle contraction. Between 50 and 80 volts, blisters commenced along the 5-cm wire contact within 20 to 30 seconds of current flow. At approximately 200 volts, the current reached its maximum value in 1 second or less, followed by arcing and current diminution. At 500 volts, full-thickness (third-degree) skin burns occurred in 1 to 2 seconds with currents in excess of 1 amp. However, no burns occurred under the large (40×60 cm) plate. When the 5-cm disk was applied with 0.1 kg/cm^2 on the hind limb opposite to the large 40×60 cm plate, the temporal pattern of current flow depended on the voltage selected. The muscles first responded with tonic (sustained) contractions. Then clonic (jerking) contractions were observed, with tissue charring and a current decrease. Burning began at the periphery of the electrodes and moved inward. Blisters formed a ring 2 to 3 mm around the electrode. The time required for current buildup, skin necrosis, and the current decrease phase was directly proportional to the electrode size.

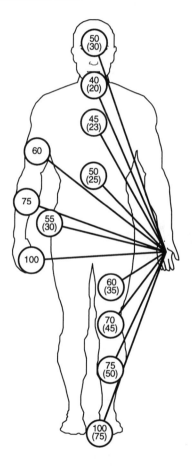

Figure 6.4. Percent impedance for large area contact between the hand to the site identified; 100% represents the hand-to-hand impedance. The bracketed numbers refer to the percent impedance when both hands are joined and current flows to the site indentified. (From report 479-1 Commission Electrotechnique International 1984, by permission.)

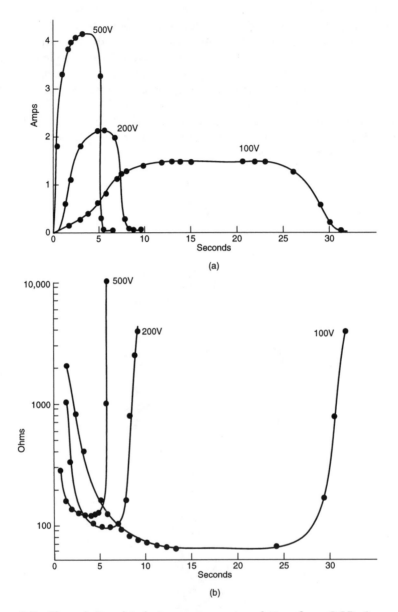

Figure 6.5. The relationship between current and time for a 6.25×1 cm elliptical electrode on hog skin, paired with a 40×60 cm electrode for the application of 500, 200, and 100 volts (A). For each measurement the electrode was
moved to a clean site. In B are shown the impedance versus time relationships
for the same electrodes and voltages. (Redrawn from data published by Sances
et al., IEEE Trans. Appar. Power Systems 1981, PAS 100 (6): 2987–2992).

The rapid nature of the current rise and fall depended on the initial voltage, as shown in Figure 6.5A for a 6.25×1 cm elliptical electrode paired with the 40×60 cm electrode on the opposite hindquarter. In Figure 6.5B is shown how the impedance changed with time. In all cases, the impedance decreased, reached a minimum and then increased markedly. With current flow the tissues became hot and the impedance decreased. When boiling and arcing occurred the impedance increased dramatically.

Sances et al. (1981) also measured the manner in which the current is a function of the increase in applied voltage and the electrode area. Using disks of 0.5, 1, 2.5 and 5 cm diameter, paired with the large 40×60 cm plate under the opposite hindquarter, they varied the voltage and measured the current flow for each electrode size; Figure 6.6A presents the results. In all cases, the current increased then decreased and increased again with increasing voltage. Figure 6.6B shows the impedance-voltage relationships for all electrodes. The initial impedance decreased with increasing voltage is likely due to the rise in temperature of the tissues as well as breakdown of the skin dielectric. At a critical point, the impedance rises, probably at the point of tissue boiling and arcing. Beyond this point, the impedance decreases slightly with increasing voltage (and current), probably due to bulk tissue heating.

The foregoing clearly shows that although the impedance of the electrode-subject circuit decreases with increasing area of contact, it also decreases with increasing voltage, up to about 100 volts. Beyond this value, the impedance rises, but is still dependent on electrode area. Therefore it is clear that although the bulk of the body may have a reasonably identifiable impedance, the dynamic events at the site of contact between the skin and a conductor have a profound effect on the current flow.

CURRENT DISTRIBUTION

As stated previously, current seeks the lowest resistance pathway. In a stratified conductor such as the living body, the conducting properties of the various tissues and organs are different. Resistivity is that property of a material to oppose the passage of current and will determine the current distribution. Therefore, a more accurate statement for a stratified conductor is "Current seeks the path of lowest resistivity". Because the heating is proportional to the product of time, current density squared and resistivity, the most heating will occur in tissues of lowest resistivity. Postmortem examination of legally electrocuted criminals provides information that is consistent with the concept that the current will seek the lowest resistivity tissues and heating in such tissues will be higher than in tissues with higher resistivity. For example, the histological

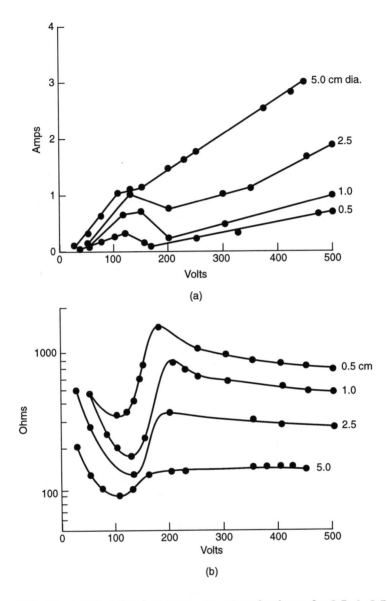

Figure 6.6. The relationship between current and voltage for 0.5, 1, 2.5 and 5 cm disk electrodes on hog skin, paired with a 40×60 cm electrode under the animal (A). In B are shown the corresponding impedance-voltage relationships. (Redrawn from data published by Sances, A.D. et al. IEEE Trans. Power Appar. Systems 1981, PAS 100 (6): 2987–2992.)

changes in brain tissue were reported by Hassin (1933) who examined the brains of five criminals electrocuted with alternating current. Swelling, tearing, and liquefaction of cells were found. In some instances the larger blood vessels were ruptured. The temperature at different body sites following legal electrocution was reported by Werner (1923), who measured a temperature of 120°F on the right leg, 145°F in the brain, and a slightly lower temperature in the lungs and abdomen; the blood was hemolyzed. The heart was not beating, but it could be stimulated to contract. An indication that current may flow preferentially along blood vessels and nerve fibers was presented by Spitzka and Radash (1912), who reported tearing of vessel walls. They stated that "the maximum number of lesions are found in the most constricted part of the brain stem in the path of the current and are most numerous along the longitudinal fiber tracts and blood vessels."

Sances et al. (1983) carried out power-line current-distribution studies in the hog with forelimb-to-hindlimb electrodes (copper wire wrapped around the distal members). Strong muscle contractions were obtained with 15 volts (30 mA). They measured current density values at many sites using a tetrapolar probe to measure voltage gradient and resistivity. They found, that for unit area, nerves and blood vessels carried the most current. With 10 amps flowing for 30 sec, the temperature in the spinal cord rose to 42°C. Muscle, fat and bone carried less current. With 2,000 volts, arcing was routinely encountered at the wire electrodes. In no case was the phase angle between the voltage and current more than 5 degrees, indicating that the circuit was mainly resistive. This and the previous studies clearly indicates that there are preferential pathways for current flow.

REFERENCES

Baker, L.E. and Mistry, G.D. Assessment of cardiac function by electrical impedance. Proc. 7th ICEBI, 1981, Tokyo, pp. 7–10.

Bozler, E. and Cole, K.S. Electrical impedance and phase angle of muscle in rigor. J. Cell. Comp. Physiol. 1935, 6:229–241.

Burger, H.C. and vanMilaan, J.B. Measurement of the specific resistance of the human body to direct current. Acta Med Scand 1943;114:584–607.

Burger, H.C. and vanDongen, R. Specific electric resistance of body tissues. Physics Med Biol 1960–61;5:431–447.

Cole, K.S. and Curtis, H.J. Electrical impedance of nerve and muscle. Cold Spring Harbor Symp. Quant. Biol. 1936, 4:73–89.

Cole, K.S. and Guttmann, R.M. Electrical impedance of the frog egg. J. Gen. Physiol. 1942, 25:765–775.

Commission Electrotechnique Internationale (Report 479-1, 1984). 3 rue de Varembe, Geneve, Suisse.

Crile, G.W., Hosmer, H.R. and Rowland, A.F. The electric conductivity of animal tissues under normal and pathological conditions. Amer. J. Physiol. 1922;60:59–106.

Coster, H.G.L. and Zimmerman, U. The mechanism of electric breakdown in the membranes of valonia utricularis. J. Membr. Biol. 1975, 22:73–90.

Coster, H.G.L. and Zimmerman, U. Dielectric breakdown in the membranes of valonia utricularis. Biochim. Biophys. Acta 1975;382:410–418.

DeGaravilla, L., Tacker, W.A., Geddes, L.A., et al. In-vitro resistivity of canine heart to defibrillator shocks. Proc. 1981 AAMI 16th Ann. Mtg. p.28.

Freygang, W.H. and Landau, W.M. Some relations between resistivity and electrical activity in the cerebral cortex of the cat. J. Cell. Comp. Physiol. 1955;45:-377–392.

Fricke, H. The electric conductivity and capacity of disperse systems. Physics 1931, 1:106–115.

Fricke, H. and Curtis, H.J. Electric impedance of suspensions of leukocytes. Nature. 1935, 135:436.

Fricke, H. and Curtis, H.J. Specific resistance of the interior of the red blood corpuscle. Nature. 1934, 133:651.

Galeotti, G. Uber dfie elektrische Lietfahigkeit der tierschen Gewebe. Z. Biol. 1902;43:289–340.

Gauger, B. and Bentrup. F.W. Study of dielectric breakdown. Membr. Biol. 1979;48:249–264.

Geddes, L.A. and Baker, L.E. The specific resistance of biological material. Med. Biol. Eng. & Comput. 1967;5(3):271–293.

Geddes, L.A. and Sadler, C. The specific resistance of blood at body temperature. Med. Biol. Eng. 1973;11(3):336–339.

Hassin, G.B. Changes in the brain in legal electrocution. Arch. Neurol. Psychiat. 1933, 30:1046–1060.

Hemingway, A. and McLendon, J.F. The high frequency resistance of human tissue. Amer. J. Physiol. 1932;102,56–59.

Kanai, H., Sakamoto, K. and Miki, M. Impedance of blood; the effects of red cell orientation. Dig. 11th Int. Conf. Med. Biol. Eng. pp. 238–239, 1976.

Kaufman, W. and Johnston, F.D. The electrical conductivity of the tissues near the heart and its bearing on the distribution of the cardiac action currents. Amer. Heart J. 1943;26:42–54.

Kinnen, E., Kubicek, W., Hill, P. and Turton, G. Thoracic cage impedance measurements. (Tissue resistivity in vivo and transthoracic impedance at 100 kc/s) Tech. Doc. Rep. SAM-TDR 65-5. 1964. School of Aerospace Medicine. Brooks AFB, Texas.

Lepeschkin, E. Modern Electrocardiography, Vol. 1. Willams and Wilkins, Baltimore 1951.

Liebman, R.M. and Cozenza, F. Study of blood flow in the dental pulp by an electrical impedance technique. Phys. Biol. Med. 1962–1963;7:167–176.

Moskalenko, Y.E. and Naumenko, A.I. Movement of the blood and changes in its electrical conductivity. Bull. Exp. Biol. Med. 1959;47:211–215.

Nicholson, P.W. Specific impedance of cerebral white matter. Exp. Neurol. 1965, 13,386–401.

Osswald, K. Messung der Leitfahigkeit und Dielektrizitatkonstante biologischer Gewebe und Flussigkeiten bei kurzen Wellen. Hochfreq. Tech. Elektroakust. 1937;49,40–49.

Radvan-Ziemnowicz, J.C., McWilliams, J.C. and Kucharski, W.E. Conductivity versus frequency in human and feline cerebrospinal fluid. Proc. 17th Ann. Conf. in Med & Biol. (1964) McGregor and Werner, Wash, 12, D.C.

Ranck, J.B. Specific impedance of rabbit cerebral cortex. Exp. Neurol. 1963;7:1-44–152.

Ranck, J.B. and Be Ment, S.L. The specific impedance of the dorsal columns of cat; an anisotropic medium. Exp. Neurol. 1965;11:451–463.

Rosenthal, R.L. and Tobias, C.W. Measurement of the electric resistance of human blood using coagulation studies and cell volume determination. J. Lab. Clin. Med. 1948;33:1110–1122.

Rush, S., Abildskov, J.A. and McFee, R. Resistivity of body tissues at low frequencies. Circulation Res. 1963;12:40–50.

Sakamoto, K. and Kanai, H. Electrical characteristics of flowing blood. IEEE Trans. Biomed. Eng. 1979 BME 20:687–695.

Sances, A.D., Szablya, J.F., Morgan, J.D. et al. High voltage powerline injury studies. IEEE Trans. Power Appar. Systems 1981, PAS100: 552–558.

Sances, A.D., Mykelbust, J.F., Larson, S.J. et al. Experimental electrical injury studies. J. Trauma 1981, 2(8): 589–597.

Sances, A.D., Mykelbust, J.B., Szablya, J.F. et al. Current pathways in high-voltage injuries. IEEE Trans. Biomed. Eng. 1983, BME 30:118–124.

Schwan, H.P. and Li, K. Capacity and conductivity of body tissues at ultrahigh frequencies. Proc. IRE 1955;41:1735–1740.

Schwan, H.P. Electrical properties of body tissues and impedance plethysmography. IRE Trans. Med. Electron. PGME 1955;3:32–45.

Schwan, H.P. and Kay, C.F. The conductivity of living tissues. Ann. N.Y. Acad. Sci. 1956–57;65:1007–1013.

Schwan, H.P. and Kay, C.F. Specific resistance of body tissues. Circulation Res. 1956;4:664–670.

Schwan, H.P., Kay, C.F., Bothwell, P.T. and Foltz, E. Electrical resistivity of living body tissues at low frequencies. Fed. Proc. 1965;13:131.

Schwan, H.P., Kay, C.F., Bothwell, P.T. and Foltz, E. (1965) Electrical resistivity of living body tissues at low frequencies. Fedn Proc. Fedn Am. Socs exp. Biol. 13, 131.

Sigman, E., Kolin, A., Katz, L.N. and Jochim, K. Effect of motion on the electrical conductivity of the blood. Amer. J. Physiol. 1937;118:708–719.

Spitzka, E.A. and Radash, H.E. The brain lesions produced by electricity as observed after legal electrocution. Am. J. Med. Sci. 1912, 144:341–347.

Tacker, W.A., DeGaravilla, L., Babbs, C.F. et al. Resistivity of blood to defibrilator-strength shocks. Proc. 17th Ann. Mtg., 1982, AAMI p. 121.

Tacker, W.A., DeGaravilla, L. Babbs, C.F. et al. Resistivity of blood to defibrillation-strength shocks. Proc. AAMI 17th Ann Mtg. 1982, San Francisco, page 121.

Tacker, W.A., Mercer, J., Foley, P. and Cuppy, S. Resistivity of skeletal muscle, skin, fat and lung to defibrillation shocks. Proc. AAMI 19th Ann. Mtg. Apr 14–18, 1984, page 81.

Van Harreveld, A., Murphy, T. and Nobel, K.W. Specific impedance of rabbit's cortical tissue. Amer. J. Physiol. 1963;205:203–207.

Velick, S. and Gorin, M. The electrical conductance of suspensions of ellipsoids and its relation to the study of avian erythrocytes. J. Gen. Physiol. 1940;23:-753–771.

Werner, A.H. Death by electricity. Med. J. Med. Record 1923, 118:498–500.

INDEX